Managing the Digital You

Managing the Digital You

Where and How to Keep and Organize Your Digital Life

Melody Condron

ROWMAN & LITTLEFIELD
Lanham • Boulder • New York • London

Published by Rowman & Littlefield
A wholly owned subsidiary of The Rowman & Littlefield Publishing Group, Inc.
4501 Forbes Boulevard, Suite 200, Lanham, Maryland 20706
www.rowman.com

6 Tinworth Street, London SE11 5AL, United Kingdom

Managing the Digital You: Where and How to Keep and Organize Your Digital Life is part of the Rowman & Littlefield LITA Guides series. For more information, see https://rowman.com/Action/SERIES/RL/LITA.

British Library Cataloguing in Publication Information Available

Library of Congress Cataloging-in-Publication Data

Names: Condron, Melody, 1979- author.
Title: Managing the digital you : where and how to keep and organize your digital life / Melody Condron.
Description: Lanham : Rowman & Littlefield, [2017] | Includes bibliographical references and index.
Identifiers: LCCN 2016047505 (print) | LCCN 2017003918 (ebook) | ISBN 9781442278875 (cloth : alkaline paper) | ISBN 9781538126233 (pbk. : alk. paper) | ISBN 9781442278882 (electronic)
Subjects: LCSH: Electronic records—Management. | Personal archives—Management. | Digital media. | Digital communications.
Classification: LCC CD974.4 .C65 2017 (print) | LCC CD974.4 (ebook) | DDC 070.5/797—dc23
LC record available at https://lccn.loc.gov/2016047505

Printed in the United States of America

Contents

Preface

Things have changed. Just 20 years ago, managing your personal files may have consisted of keeping medical, legal, and personal papers in boxes or files for later access. Emergencies like flood or fire may lead to loss, but in general, files kept in a box or folder stayed where they were until you needed to use them. If you wanted to be extra vigilant, you could make copies or store them in a safe deposit box at the bank. Many people still follow these practices, but it is more difficult to maintain a fully physical system in the highly digitized world. Documents like licenses and deeds continue to be physical, but insurance paperwork, bank documents, and bills are increasingly digital. Personal correspondence, too, is predominantly email and SMS/text. Yet few of us are confident in our digital systems, and we are not backing things up. This leads to potential losses, not just through accident or theft, but also because of the known limited life of hardware and the inaccessibility of old software over time. Google Vice President Vint Cerf recently warned that a generation, or even a whole century, is at risk of being "forgotten" as the software and hardware holding with our communal and personal histories get older and die.[1] This risk has birthed an interest in personal digital archiving.

Perhaps the largest challenge for all computer users is dealing with the volume. Virtually every service and tool requires an account for full access. With these multiple accounts come multiple passwords that we are not supposed to write down or share. We are also creating our own digital items faster than ever before. Blogger Benedict Evans calculates that over 2 trillion photographs were shared online in 2015—a number larger than the estimate of all photos taken on physical film, *ever*.[2] We additionally have hardware that fails as it gets older, software that changes with every update, and files that can become corrupted or lost in the large number of poorly named items on our hard drives.

So how do we manage this volume and chaos? How do we manage our multitude of accounts, overflowing emails, and digital items spread out across multiple devices and websites? People cope with the overwhelming digital environment in different ways, depending on their familiarity with computers and personal experiences. Generally, users learn to cope by coming up with custom solutions: things that work for them but may not make sense to anyone else. This may work for a while, but it ultimately means that no one can help you if you mix up something or if someone else needs to know how to manage your documents or accounts (such as in an emergency). Another common coping mechanism is to avoid digital options whenever possible. This isn't practical as things increasingly move online, creating a more difficult situation for people who eventually hit a situation that forces them to the computer. Finally, many computer users just never develop a system at all, or are inconsistent.

The truth is that there is no one "best way" to manage your files, because each of us have different needs and expectations. No one solution exists to address the common problems of having too many accounts, too many online systems, and too many digital documents. Instead there are many possible solutions that have worked for others, and are based on logic as well as trial and error. One of them may work for you.

Managing the Digital You can help. It will offer basic and practical solutions that are attainable and recommended by others with similar needs. It will help you identify and prioritize. Finally, it will offer tips and tools to get you started, even if you are not a computer expert and have limited funds.

HOW DID THIS HAPPEN

You may wonder, "How did things get so difficult? And how did we acquire so many digital files and photos so quickly? Where did all this stuff come from?!" Part of the issue is that things have changed slowly, and most people have not adapted. The transformation happened over a matter of years, with small incremental changes that made it easy to ignore for a while. Then, suddenly, like a lobster in water that's been slowly getting hotter, we can't ignore it any longer. We reach a tipping point where our old systems that worked for a while just don't work anymore: we can't find anything, everything we try to do is frustrating, and we don't know how to fix it. It has gotten complicated in small parts, so that it is hard to go back and rebuild without losing things. Even more frustrating, you can't just scrap your computer and start over without losing photos and files you have tied up in your existing systems. Whether you embraced computers early on, or only use them when you must, there is no one silver bullet to help users grapple with the overwhelming number of options. There are thousands of applications available to organize photographs, for example. How do users assess so many tools? Where do they start? What can be done when there are new photos, documents, and emails being created every day on top of a backlog of items already spread out across multiple devices? *Managing the Digital You* will hopefully serve as an answer to these questions by giving users the basic tools they need to get started with personal digital archiving.

REASONS TO PLAN FOR YOUR DIGITAL ITEMS

Beyond the desire to get ahead of the mounting digital landslide, there are a number of good reasons to organize and manage your digital items in a meaningful way. You probably know many of them already, since you have picked up this book. While everyone will have their own personal reasons for wanting to manage and save their digital items, a few common themes emerge.

Avoid Loss

Even if users are not overwhelmed by their digital stuff at this point in time, creating a backup system and planning for contingencies can stave off difficult situations later. The impetus to preserve and protect often comes after a significant loss, but that does not have to be the case. Individuals who have experienced a hard drive crash, laptop theft, or a home disaster like a fire obviously understand the importance of protecting items after the fact—but by then the damage may be done. The risk is much higher than some would believe. It is not just accidental and chance loss that threatens digital items: the systems themselves are built to fail. Hard drives and flash drives have a limited life, which can be drastically shortened depending on environmental conditions like extreme heat. Laptops and storage devices are at added risk because they are exposed to more environmental extremes, as well as being more likely to get dropped, stolen, or otherwise ruined.

Even when considering a desktop computer at low risk of theft, having a single device responsible for protecting all of your digital items is simply not good planning. Computer crashes and drive failure are common. Creating a backup system that can reduce the likelihood of loss is common sense, and is easier than many people expect. Working hand in hand with backups, deleting unused and unwanted files and naming others appropriately will make files easy to retrieve and backup systems more efficient overall.

Help Family and Loved Ones in Case of an Emergency

Like having a will, good personal digital archiving is something everyone can do to make it easier on their family in cases of emergency or death. Consider that any bank accounts, bills, investments, or insurance policies stored online are likely impossible for family members to access, if they even know that the accounts exist. If computer and email accounts are password-protected, even the most savvy computer users won't be able to find those accounts in time of need. While it isn't a good practice to hand out passwords or keep them on a sticky note next to the computer, there are a number of other options that allow some access to family members in case of emergency. Some of these options will be discussed in the section on legacy planning. Users with family photos and similar shareable items may also want to focus on digital archiving in order to share what they have with friends and family.

Find and Share with Ease

Whether sharing photos so that they will never disappear, or just sharing vacation photos for fun on Facebook, organizing items makes sharing easier. It is difficult to retrieve photos when they all have batch names assigned by a camera or import program, usually consisting of long number-and-letter strings. It can also be challenging to find and share items when they are kept in multiple different places. For example, many people have photos on their phone, tablet, digital camera, computer, and online. Remembering where items are stored can become next to impossible as a collection grows. Since the goal of most amateur and family photographers is to share items and look back at old photos later, managing these items in a meaningful way is important.

The same is true of documents, not just photographs. One common issue when working with documents is the creation of multiple versions of the same document in draft forms. This may happen because multiple people are working on a document together. Applying some basic naming conventions can help avoid confusion, and keeping things in a consistent manner will make them easier to find again later.

Make Life Simpler

In a 2011 study, researchers looked into whether creating email folders or other forms of email organization were worth the time spent to set up and maintain that sort of organization system. Despite a large percentage of email users depending on folder organization, the study found that email searching did a better job with less frustration.[3] Search mechanisms have only improved since then, and many users may want to consider this as a lower-maintenance method of managing their inbox.

This is just one way that value assessment comes into play in personal digital archiving. By assessing your digital stuff, identifying what values you have, and making plans for only the important items, you can drastically simplify your life. In addition, you can cut the ongoing maintenance time needed to keep your computer and files in good working order. For some users, folder maintenance is easy to maintain or, if difficult, is worth the effort. Other users, however, need to look at the time cost of maintaining a folder storage system. Does the organization system really help, or is the built-in search function just as good? Do they need to store emails in folders for easy backup, or do they not care about backing up email? A similar assessment may be needed for other digital items. For example, an individual may be scanning and keeping digital copies of utility bills because they think they should. However, when asked what they plan to do with them or what they might use them for in the future, they might discover that they have no intended purpose for those documents.

Clean Up Physical Items

While we may never reach the paperless office promised when computers first became popular, many of the paper documents we want to keep could be more efficiently stored on our computers. Digital copies of important documents are nice to have as a backup, and other items may become disposable once a digital copy is saved and backed up. This of course means that offices filled with paper might be reduced to a more comfortable workspace where most files are digital. Obviously, just digitizing files will not help you much if you do not plan for a way to store and save them so that you will have access to them as needed.

HOW THIS BOOK CAN HELP

Managing the Digital You can help you with all of these things. After reading this book, you should:

- Know where you should (and shouldn't) spend your time, based on value assessments that will help you find out what is really important to you based on your interests and needs.
- Learn about non-proprietary solutions wherever possible, so that you are not tied to a program or system that disappears. This will protect you and your files in the long run.
- Be aware of multiple strategies for managing digital problems, since not everyone wants or needs the same solutions.
- Have a basic understanding of the many options available for personal digital archiving, including options for some digital items that are often overlooked.

Each chapter is set up to cover an important topic in personal digital archiving, with the earlier chapters laying a foundation for the others. Consider reading chapters 1 and 2 regardless of your personal interests. The later chapters can be skipped or read as needed, depending on the interests of the user. Though users may be computer-savvy, the focus will be on understanding the basics of how individuals can create a system that works for them personally. Numerous tips are available in each section that may be new to even adept computer users.

All readers will want to start with chapter 1. It is the most important chapter in the book. While you can skip ahead and learn a lot in chapters specific to one kind of media or issue, chapter 1 offers a guided assessment of what digital items you own and what you value. Since the value assessment will help you decide what you need to preserve and what you really don't care about, starting with this assessment is the best way to begin personal digital archiving.

Chapter 2 explains best practices for file naming, file structure, and organization options. This chapter will cover practical advice that should become the basis for all other personal digital archiving projects. It does not tell you exactly what to name your files, but it helps you know what to avoid and makes some suggestions based on what others have done successfully.

Chapter 3 breaks down considerations for some of the most important digital items: financial, legal, and medical documents. This chapter will discuss the pros and cons of keeping digital and physical copies of these items. It also has a schedule of which items should be kept, whether a physical or digital item is best (and why), and how long each type of document should be retained. Beyond managing these items for your own purposes, this chapter also discusses how to plan for the future and create a legacy plan, which is similar to a will for your digital stuff.

Chapter 4 looks at correspondence including email, SMS (texting), and voicemail. This chapter will discuss what options exist for archiving digital correspondence. Since this depends heavily on what devices and software are in use, this chapter will highlight some of the major differences between email providers. In addition, current applications and tricks for capturing these types of media will be highlighted.

Chapter 5 covers best practices for the most popular personal digital archiving topic: photographs. So many resources and applications exist to help users manage their photos and multimedia that it is difficult to keep track and to choose a tool that holds up. In addition, much of the literature available on the subject of digital-photo best practices focuses on guidelines for institutions that are impractical for individuals. This chapter offers practical advice for organizing and managing multimedia so that you can find and retrieve items when you need them.

Chapter 6 focuses on other multimedia, including audio and video. It will also discuss special cases like genealogical files and outdated media. Look to this chapter if you have old hardware, old software, or files that can no longer be opened with your current computer.

Chapter 7 is all about social media, online sharing, and online accounts. This chapter will explain what aspects of social media can be easily archived, and what can be archived through more clever means. In addition to social media sharing, this chapter will look at the benefits and risks of shared folders and files on platforms such as Dropbox and Google Drive. While you are online looking at your social media accounts, this chapter can help you to review your other online accounts.

Finally, chapter 8 will look at the non-digital material: the stuff you have waiting around that you are also worried about losing. There are entire books devoted to scanning and digitizing, so this chapter will only serve as an outline of your options. It provides some of the basics for getting started on the backlog of physical stuff.

One note of caution as you get started: some things in this book may not work for you, and not just because you don't like them. Some of the directions may not work depending on your computer, your operating system, or changes to software or systems after this book was written. I urge you to be adaptable, and to always seek out more information where needed. The tools and directions in this book worked with operating systems in common use as of late 2016, but computers and software are constantly changing. Where possible, specific operating systems are mentioned in the directions to help you.

No matter what your interest is in archiving your personal digital items or how much digital stuff you have, the most important thing is to get started. Prioritize the things that are most important, back up what you have, and schedule some time to work on your plan. Finally, stay positive and give yourself a break! The digital world is not easy to wrangle, but you *can* preserve and protect what you need to with a little effort.

NOTES

1. Marc Kosciejew, "Digital Vellum and Other Cures for Bit Rot," *Information Management Journal* 49.3 (2015): 20–25.

2. Benedict Evans, "How Many Pictures?" *Benedict Evans* (blog), August 27, 2015, http://ben-evans.com/benedictevans/.

3. Steve Whittaker, Tara Matthews, Julian Ceruti, Hernan Badenes, and John Tan, "Am I Wasting My Time Organizing Email? A Study of Email Refinding," *Proceedings of the SIGCHI Conference on Human Factors in Computing* (May 2011), 3449–58, doi:10.1145/1978942.1979457.

1

Getting Started

Finding Your Files, Assessing Value, and Making a Plan

When people first start reviewing their digital items, organizing and protecting them can seem like an overwhelming prospect. Items may be spread out on multiple different devices, as well as online, and it is not immediately clear how to collect them all into one place. Deciding how to proceed involves three steps:

1. Assess what is important. Doing this first makes it clear what is worth time to organize and protect, and what is not.
2. Create a list of where all digital items are located right now. This includes storage devices, computers and tablets, and online locations.
3. Make a plan for how to effectively store and back up items. This is the most complex of the three steps, but is where systems actually come together. You will need to complete steps 1 and 2 before setting up a backup so that you know you have not missed any file locations on your backup list.

All of these steps are explained in detail in this chapter. Forms and directions will help you get started. As part of the process, you will create a plan for saving, storing, and backing up your items now and in the future. Keeping a record of how you organized things will help later if you forget, since you do not want to deal with your backup system very often. For that reason, it may help to start a word processing document with your plans. You can also write down your plans in a notebook as you set things up, which may be helpful if you don't have a great organizational system set up and are concerned that you may lose a digital file (it happens!).

STEP 1: ASSESS VALUE

Assessing personal value is a huge part of personal digital archiving. Catherine C. Marshall identifies the problem succinctly: "Digital material accumulates quickly, obscuring those items that have long-term value."[1] We are so inundated with digital items that we don't often take the time to assess them or their importance. Instead, we just let them build up because we feel overwhelmed, or don't know where to start. For example, consider someone who spends a lot of time worrying about the many online bills and statements they receive. In particular, they are concerned that they are not all downloaded and organized for later access. These items used to arrive as physical paper statements in the mail, but now are available on the company website behind a login screen because e-statements sounded easier (at the time). Usually consumers get an email telling them their statement is available. However, the statement itself is not included in the email for security purposes, so the email itself is not helpful to save. When asked directly, many people may not be able to explain why they need to have records of their utility bills. Yet they do want their medical bills and bank statements—those are crucial. Realizing that they don't need to keep everything—because not everything is valuable or helpful to them—cuts the time needed to download and organize digital statements. The same is true of email. Most people get a lot of email that isn't important to them. If there is only a handful worth keeping, those items may be the only emails worth organizing and backing up. Identifying what is the most important also helps people to prioritize: if something is crucial and you only have a little time, crucial comes first.

Use a value assessment form like the one in table 1.1 to identify what things are most important for you personally to save and back up. The form is also available in appendix A. Family photos are often high on this list for many people, but this is a personal assessment; not everyone values things the same way. You may find that protecting your medical and banking records is the most important thing because it gives you peace of mind to know that those items are safe. In any case, make note of what ranks highest on the list for you. Also take note of the things that you would not notice or care about if they suddenly became unavailable. In some cases, like email, you might have a portion that are important and others that are not. Some items may not be valuable now but could become valuable: for example, you may not care about your tax records until you get audited and cannot find them. If there are items that you legitimately don't care much about *and* you cannot think of future problems that might occur if those items were lost, those are items you do not have to waste time on.

Once you have completed the value assessment, ask yourself the following questions:

- Is there anything that I marked as highly important that is not currently backed up? These items will be your highest priority when setting up a backup system later in the chapter.

Table 1.1. Value Assessment

MY STUFF	N/A: Does not apply to me.	If I lost this forever, I might not notice or I wouldn't care.	If I lost this forever, it would be inconvenient but I'd be fine.	If I lost this forever, it would ruin my day but I would manage.	If I lost this forever, I would be sad about it for a long while.	If i lost this forever, I would be devastated.
Photos of family and friends						
Emails						
Texts/SMS						
Personal writing (journal, fiction, etc.)						
Bills or financial records						
Medical records						
Scanned letters or correspondence						
Music						
Family videos						
Legal documents						
Saved voicemails						
Genealogy research						
Other:						
Other:						
Other:						
Other:						

• Is there anything I have been keeping or spending time organizing that—when I really consider it—isn't worth my time? These are things you can stop worrying about and maintaining. When everything else is set up, you might consider deleting some of these items, depending on what they are.

STEP 2: CREATE A LOCATION LIST

Now that you know what you have and what it is worth to you, it is important to know where all your items are kept so that you can set up a backup system for these locations. You can do this in a few ways, and it helps to be thorough, since it is often difficult to think of all of the digital places items might be. Ideally this will be a list of where unique items are kept, so you may want to skip locations that are only copies. For example, if you burned CD backups of the photos stored on your computer, you don't need to list those CDs. Also, if all of the photos on your Facebook page are copies and exist somewhere else, then you don't need to list Facebook. However, if there are photos or conversations on Facebook that you want to save, and they don't exist anywhere else, then you will want to add it to your location list.

Create a list of all locations you can think of that have unique digital items:

1. Consider the items from the value assessment list that you have already created. Use a form like the one in table 1.2 to list where these items are stored. The form is also available in appendix A.
2. List all devices that you use, at home, work, and in between.
3. List all websites that you upload to or download from, including social networks, cloud/online storage, banking and bill websites, and shared drives. Places commonly overlooked include your genealogy records on a family history website, or medical websites that give you access to past tests. If these sites include items that don't exist anywhere else, then they should be included.
4. List all physical storage devices where you have items saved, including flash drives, CDs, old laptops, and external hard drives. Don't forget any outdated media that you may have stashed away in storage, such as Zip disks and floppy disks.

Some of the locations may have reminded you of items that were not on your value assessment list, and that's OK. You can add them as needed or disregard them if they aren't important to you. Once you have your location list finished, you will need to identify which locations need to be backed up and which don't. This is based on your value assessment. If you see a location that only includes things that are not important to you, you can then cross it off the list. In the case of high-priority items that are of great value, bold or highlight the locations where those items are stored. Keep the medium-priority items on the list for now. It will be easier to assess whether those items should be backed up depending on how much work and time it will take to do it.

Table 1.2. Location List

DIGITAL ITEMS	*LOCATIONS*
Photos of family and friends	
Emails	
Texts/SMS	
Personal writing (journal, fiction, etc.)	
Bills or financial records	
Medical records	
Scanned letters or correspondence	
Music	
Family videos	
Legal documents	
Saved voicemails	
Genealogy research	
Other:	
Other:	
Other:	
Other:	

STEP 3: PLAN AND BACKUP

Before moving on step 3: have you completed steps 1 and 2, listing all of your digital items and locations and whether they are important? If not, I urge you to do so before setting up your backup system. Skipping ahead may create more confusion. For one thing, you may set up your backup and realize later that some of your online files are not backed up, or that you forgot a device and now it isn't included in your plan. You can always add things, of course, but sometimes it helps to design your backup around what you use and have. The value assessment is also important. If you have multiple devices but the files on some of them are not very important to you, you may be able to make your backup plan simpler by not including them.

With a list of items and their locations, it should be much more straightforward to see what needs to be backed up. At this point it is common to feel overwhelmed, since digital items may be spread out all over. Backing up every device and location individually would indeed be difficult and time consuming. To make the backup easier, consider designating one location that is the hub for all of your digital items. By designating a hub, such as your home desktop computer, you can set up systems that regularly move your items to the hub. Then, because your items are all in one place, you only need to backup your hub location. This system may not work for everyone, but it greatly simplifies storage and backup if you can make it work. If moving some items to a hub isn't possible for whatever reason, consider at least limiting the devices that are part of your backup system to a small number. This will avoid the need to remember to manually save things. For example, you may be able to back up your laptop and phone separately but not your tablet—and that may be just fine if your value assessment shows you that you don't keep crucially important things on your tablet, anyway. The best backup system is the one that is done consistently, and most people do not remember to back up their systems. Automated systems are far more reliable than trying to remember and make time.

TIP: Dropbox and other storage providers that create a folder location on your computer may work as your hub. As long as you choose backup systems that can backup Dropbox or other online folders, then the hub will work. Since so many people use services like Dropbox to remotely store their files, this may be the best option for some users as long as they feel comfortable with their primary storage location being online instead of local. However, they must remember to store files in Dropbox rather than locally on their computer, or these items may not be backed up.

This may sound obvious, but is often overlooked: online storage is only a backup if it is a copy. For some reason, many of us have a difficult time wrapping our head around this fact. If you keep the main copy of something on Google Drive, One-Drive, CloudMe, Amazon Cloud, Dropbox, or any of the other online file storage locations, then it is not protected. Many people keep these files online for file sharing or so that they have access to them from multiple locations. However, having a single copy of anything digital is never a good idea: files get corrupted or accidentally deleted all the time—more often when multiple people have access to them. Google Drive may save versions of your file, but versions can be difficult to sift through, and deleted files are hard or impossible to recover.[2] If you use these services regularly, make sure you include them in a backup plan or use them as the backup and store your "primary" copy of a document elsewhere. It is also highly recommended that you set up your backup system before you start to move and organize your items.

Backup Considerations

Setting up backups in the beginning means that anything you do on your primary hub device will be backed up immediately while you work out the other issues. Having it in place will lessen the urgency and lower the risk of loss. There are a number of choices to make in setting up a backup system:

1. Choose your hub location or a short list of devices that need to be backed up.
2. Choose two different options for backup, with at least one of the options being physically located in a different place than your hub device (see figure 1.1). It is extremely important to have one of the backups be off-site to avoid loss. If the location where the hub device is located is compromised by fire, theft, or other

emergency, both items may be lost. A common backup system choice includes a local backup on an external hard drive and an off-site backup using cloud storage. Using two different cloud storage services may also work. If you have chosen a hub location other than a laptop or desktop computer, you will need to investigate the best choices for that device or location. Multiple options are listed in the next section.

3. Decide what items, folders, applications, and other information you will back up from your location(s). Some services will not back up applications because of their size, and you may need to pay more for storage devices or space if you need a lot of space. If it is too overwhelming to go through all of your files, that is OK, too—in fact, many systems default to backing up all documents, music, and general files. Just know that you may end up paying more money that you need to if you use a large amount of space. Also, though you may be tempted to use less space by compressing, resist the urge to do so.[3]

4. Decide how often to back up. It is highly recommended that both backup systems be automated and run daily. Some programs are set up to run regularly whenever your computer is turned on, so you may not have to choose or decide on a schedule. Others, like Windows Backup, can be set to run at intervals based on your schedule. If you prefer a manual process rather than an automated one, decide how often to back up. If you use your computer often, daily backup is recommended.

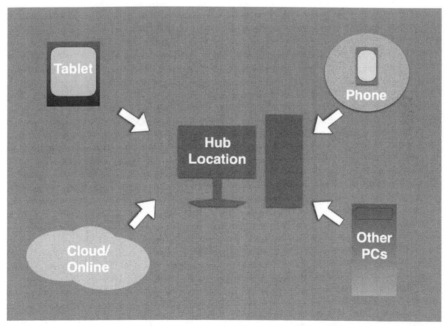

Figure 1.1. Backups work best when items from all devices are sent to a central location, or hub.

Really, Two Backups?

You may ask: do I really need two backups beyond my original? The answer to that is definitely yes. One backup is better than none, so if you can't commit to two backups for some reason, don't give up completely. Two backups, however, are preferred and recommended to avoid data loss. Consider that you have an automatic backup set up and you (hopefully) don't have to think that much about it most of the time. In fact, the only time you will probably think of it is when something goes wrong, which won't be often. If something has gone wrong and you don't check your backup system very often to ensure that files all look correct (and few people do), then you won't know something is wrong until your original copy/system fails. Then, suddenly, both your original and your backup are gone and your files are lost. A system with two backups is statistically safer, and if your two backups are different (such as an external hard drive and an online backup system), there are additional benefits. In the case of temporary Internet loss or an online backup website going down, a local backup will mean that you are never without a copy of your files if your computer fails. Similarly, if your house burns down and you lose your computer and your external hard drive backup, then your files are still safe in your online backup system. A two-backup system is a safe choice.

TIP: If you store *items in Dropbox, Google Drive, or similar online folder systems, those items are not your backups. This is a common misconception. If you keep and update files from that location, those are your originals, and should be backed up elsewhere.*

Choose Your Backup Systems

There are a multitude of options for backing up your digital files, and you can easily find resources online that explain best practices for setting up an up-to-the-second backup suitable for businesses and archives. However, there are really only a few kinds of backups that make sense for individual users, due to time and cost. Individuals who want the added protection, have special needs, and are willing to spend more time and money can look at the resources listed in appendix B to see whether more in-depth systems make sense for their particular situation. In general, though, most users need a simple system that works for them and does not cost a lot of money. Options break down to two different kinds of storage: on-site hardware, and off-site/online "cloud" storage.

Automation Is Best

Automating backups means less time working and reduced likelihood of forgetting to back things up. It is also a lot more likely that files will be effectively backed up and up to date if backups are automatic, since you can set up a plan thoroughly at the beginning and not worry about it later. Most computers now have built-in programs that allow for easy setup of one backup system, usually an on-site backup.

This is ideal for backing up to an external hard drive, network drive, flash drive, or second computer. If the standard built-in programs do not meet your needs, or your computer does not have such a program, consider purchasing one of the many software applications that offer additional features and control. Examples of these applications include Acronis Backup and Recovery, Rebit, Norton Ghost, and ShadowProtect Desktop. If you choose not to have any local backups and instead use cloud backup systems, you can expect the program to back up your files every time your computer is connected to the Internet. In all cases, the first backup will usually take longer, and your computer will run slowly while all of the files are backed up. After the initial backup, however, only new or changed files will back up, and this should not take long. Most systems have the option to pause or stop the backup if it is making your computer run slowly while you are using it for other things. Before reviewing individual device and storage options, you should check to see whether you have a built-in application. On recent Apple computers with current operating systems, Time Machine is standard. On most Windows machines, Backup and Restore or File Recovery offers a similar product.

Apple Time Machine

For Apple/Mac computers, a built-in program called Time Machine helps users set up a local backup. Time Machine can be accessed by going to System Preferences and then Time Machine (see figure 1.2). From there, users can select the

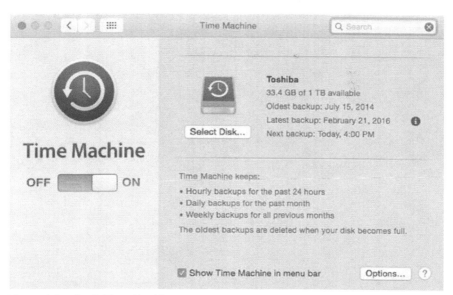

Figure 1.2. Apple Time Machine setup screen

backup disk, including Time Capsule and devices connected by USB, FireWire, and Thunderbolt. Most commonly this includes an external hard drive, but could also include a network drive or large storage flash drive. Users can choose to encrypt the backed-up data and choose files to include or exclude from the same screen. By default, Dropbox and other network folder locations are included in this backup, which is helpful for people who use these services. Once Time Machine is turned on, it keeps hourly backups for the past 24 hours, daily backups for the previous month, and weekly backups for all previous months. When your storage device becomes full, Time Mahine deletes the oldest backup. If having a long history of backups is important to you, then setting up Time Machine with a large external storage device will allow you to achieve your goals. Restoring or pulling out individual files from Time Machine is quite easy, and multiple tutorials can be found at Apple's website or through a general web search.

Windows Backup and Restore/File Recovery

Depending on your version of Windows, you may have a built-in program for backup (see figure 1.3). In Windows 10, click on the Windows logo key + s, or use the search box in the Start menu to search for/select Backup and Restore (Windows 7). In Windows 8, click on the Windows key + w and search for/select Windows 7 File Recovery. In Windows 7, select the Windows key and search for/select Backup and Restore. Any of these will bring up the Backup and Restore options window. Select Set Up Backup to select an available drive, files to backup, and scheduled

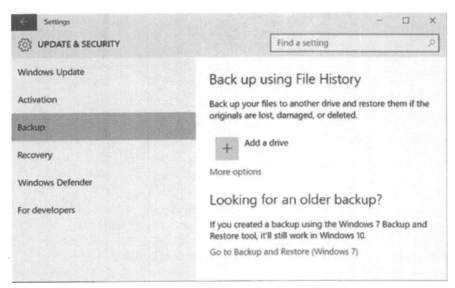

Figure 1.3. Windows backup setup screen

backup time. Unlike Time Machine, Backup and Restore does not back up regularly on its own. Instead, you have to choose a specific time and day, which can be problematic if you forget. For backups to run efficiently, the computer needs to be turned on with the backup disk attached at the scheduled time and day, or files will not get backed up regularly.

Location and Hardware Considerations

Choosing what devices and services you will need depends on your goals and individual needs. Different types of media and hardware present unique concerns and offer different benefits. *Consumer Reports* offers a computer backup system buying guide that breaks down the options and considerations nicely. Suggested considerations include: ease of use, storage size, and computer know-how, among others.[4] This chapter will give you an overview of some of the most common media locations and backup systems.

Cloud Storage/Online Backup

"Cloud" storage is data that is stored through an online/Internet backup. That does not mean your files are not actually stored in a physical place—much the opposite. Cloud storage items are backed up on numerous different physical servers, often in multiple locations. Because they are spread out and files are managed and accessed online instead of locally, they are considered to be more ephemeral and floating on the Internet: "in the cloud." Companies that offer storage and other services manage the servers, sometimes with basic services and limited space offered for free. Most of these services do not encrypt data as the default, which does mean that your files could be hacked. However, many cloud storage providers offer encryption of files as an up-charge service. If your data isn't sensitive, hacking should not be the end of the world if you have a backup. The same is true of losing a cloud storage company: if they go out of business, suddenly or otherwise, you should be safe with a backup system in place (although you will obviously need to adapt your backup system).[5]

Backing up to online storage is a good choice for one or both of your backup choices because it is relatively cheap and there are many options in the current market. The number of cloud storage options is growing, but not all cloud storage services are created equal. Some services are meant to work like a folder on your computer that you can open and save things to regularly. They are ideal for sharing items with other people and between devices. These include Dropbox, SpiderOak, Wuala, Tresorit, Meda, Google Drive, and others. Most of these offer a basic free package with options to upgrade for additional space and features. However, they do not offer ways to automatically back up your files. Instead, the idea is that you *store* (rather than back up) files. Using these cloud storage options as a backup system would require manually copying over files regularly. Other cloud storage services still give you online access to your files if you need it, but are primarily meant as a backup service. These include Carbonite, CrashPlan, Backblaze, iDrive, and others. These services cost $50–$60 each year, but they offer an automated backup service that is worth the cost. The cost often includes unlimited storage and, because it is automatic, requires no upkeep. Services like Carbonite regularly email updates and let you know if updates are not happening consistently.

TIP: Make sure that your online files are included in your backup. Since they are not backups themselves, the files you keep on service sites like Dropbox can (and should) be included in your automated backup system. Otherwise, they are at risk.

Apple iCloud

Apple iCloud is cloud storage, but it is different enough from other services to mention it in its own category. If you use only Apple/Mac products and devices, then using the Apple iCloud may work for you instead of consolidating the items on those devices. It is something of a mix between the free cloud storage used to make file sharing easier and the paid backup systems. iCloud is free for the smallest storage option, but it is very easy to go beyond the five-gigabyte limit if you allow all of your Apple devices to sync with iCloud. As with other backups, iCloud users should back up high-priority items a second time using something other than iCloud—and that is sometimes difficult, because iCloud makes it easier for people to keep items split across many devices. People who use some Apple products and some non-Apple products may find it easier to use something other than iCloud because the system is built to interact smoothly with Apple products, though it does offer service to Windows PCs. Many online tutorials exist for getting iCloud set up on a PC.[6] If Apple iCloud is part of your double-backup system, pay close attention to how you set it up and make sure that all of your items are included when you set up a second backup.

External/Portable Hard Drives

External hard drives are probably the most common attached storage and backup option. They are relatively cheap, easy to set up, and small enough to be taken on trips. This means you could back up laptops from the road or easily transfer your data to a new computer or location with ease. Built-in software on both Apple and PC machines makes it easy to create automatic backups to an external hard drive. However, there are a few considerations when choosing an external hard drive as your backup. First, they do have a limited lifetime. Backblaze, a digital storage company, runs regular tests on most of the major external hard drives on the market and reports failure rates quarterly. Their 2015 tests indicate that about 80 percent of drives last longer than four years, and failure increases exponentially after that.[7] That number is improving as technology gets better, but knowing that one in five drives fails after a few years should make any hard drive user pay attention. Having two backups and replacing any external hard drive every few years makes sense. You can purchase an external hard drive with two terabytes of space for $100–$200.[8] That is more than many people need, but most backup systems allow you to store multiple backups. That can be helpful if you don't realize that you need a deleted file for a week or two after it is gone.

Second Computer

Using a second computer as a backup is possible if you have a home network set up, but this method is usually more time-consuming to establish than an external drive or cloud backup. In addition, it requires more technical know-how to set up advanced sharing features over your home network. Each operating system has different settings and directions, so using a second computer as a backup is not highly recommended unless you are up for the extra work. As with hard drives, a second backup that is not in the same location would be important to avoid file loss from theft, fire, flood, or other calamity.

Network Drives

Like backing up to a second computer, network drives are set up through a router on your home network, making it possible to back up and share files from multiple computers and devices. Setting up network drives also requires some knowledge of network technology. *Consumer Reports* suggests that "setting up a network drive can be difficult if you're not comfortable with terms like WAP, IP address, MAC address, and workgroup."[9] If you need to back up multiple computers on a network, this option may be worth pursuing, but is likely more difficult than most individual computer users can handle.

CD/DVD

CD and DVD storage became popular when home computers started coming standard with internal burners. Many people got into the habit of saving important files to a CD and storing the disks in a safe deposit box. Certainly CDs and DVDs are still useful for sharing, but they are not a great option for backup. In fact, digital files stored on this media should be moved to your hub to protect them as part of the double-backup system. Optical disks are vulnerable to environmental conditions that can cause the storage layer to flake over time, resulting in data loss. They are also easily scratched, and are not a great way to have consistent backups since new disks would have to be burned all the time. Finally, it isn't practical to use CDs and DVDs as backups because these are becoming less common and may be more difficult to access in the future.

Flash Drive/Jump Drive

Flash drives are most useful for short-term storage, sharing, and moving items from location to location. They have only recently gotten large enough to work as backup devices, but for that purpose they are much more expensive than external drives.[10] Because they are easily lost and more susceptible to damage or failure, they are not the best backup option when compared to portable external hard drives. Despite these downsides, flash drives could be used as a backup option for short periods or if it is the only medium available. In other words, if it is all you have to work with, then certainly use the flash drive instead of not having any backup at all. This is sometimes the case for individual files of high importance in emergencies; you may not have access to your normal backup or computer, so backing up to multiple flash drives can protect individual files until a better option is available.

Backup Overview

There are many backup options available, and it can be difficult to choose. Remember, you can always adapt your system later, so don't get so mired in the difficulty that you don't actually get your backup in place. Remember these key things about backups:

- It is best to have two or more backups, with at least one being off-site.
- Setting up automatic backups reduces work and effort in the long run.
- Backup choices can include local hardware, cloud/Internet storage, or both.
- Avoid media that is not intended for long-term storage (like CDs and flash drives).
- Storing your primary documents in the cloud is NOT safe unless there is a backup in a second location.

Move It: Getting Things from There to Here

Having a backup system from your hub location is all well and good, but if you have items spread out among multiple devices, like most people, then you are still not protecting those items until they are on your hub, too. Items must be moved to the hub, ideally as part of an automatic process. Alternately, if items are in locations that you plan to stop using or have a limited number of digital items stored, then you might move things once and never have to go through the process again. Getting items moved to a hub location is tricky because everyone has different devices and locations to work with. Some of the more complex situations—like audio recordings, email, social media, voicemails, and music—will be addressed in future chapters. This section will briefly cover the basic strategies for moving items to the hub location for backup.

iPads, iPods, and iPhones

Applications for Apple devices often offer ways to sync or back up application data to iCloud, Dropbox, or other storage locations. If you add an application for your selected cloud storage option, your device may let you back up data—including photos, files, and audio. Photos, for example, can be automatically stored as a backup in Dropbox, which can then be backed up as part of your computer backup system. Most Apple devices can also be set up to back up wirelessly or through iTunes.

Tablets and e-readers

Tablets and e-readers are all very different. Check the device setup manual or find a manual online to see if the device allows backup or download of data. In the case of e-books, Nooks and many other devices purposely make it difficult to back up data because of licensing controls intended to protect copyright. Books saved as PDFs and other open sources may be easier to save and transfer. Normally any backup and transfer would be achieved by plugging the device into a computer.

Smart Phones

So many applications exist for smart phones that almost anything you want to back up from your phone is possible. Because these applications change and show up daily, it is impossible to list them here. However, one tip is to look for applications that have an option for syncing or backup to Dropbox. If you don't use Dropbox, search for applications that use your preferred system. Reading reviews in the App Store before purchase should list syncing, export, or backup as a feature.

Older Mobile Phones

Older mobile phones are much less likely to have automatic syncing capabilities and are notoriously difficult to back up. If the phone has a micro-SD card, that is often the best option for saving any photos or audio stored on the device. Micro-SD cards can fit into an adapter that can be plugged into computers with a full-size SD slot, and then items can be moved just as if they are on a flash drive.

Legacy Media and Files

During value assessment and the creation of a location list, you may come across outdated media or files. That includes old devices or storage media like floppy disks, old mobile phones, Zip disks, and basically anything that isn't something you can currently access. Even if you don't have old devices or hardware, some of your files may be unreadable or in formats that need an old version of software you no longer have. If the items are of high importance or you think they might be (but aren't sure because you cannot access them), then they will need to be converted or saved and moved to your primary hub location. Because they are on older storage media, these should be transferred to more modern technology as soon as possible.

There are two options available for getting files from old devices and storage media: do it yourself, or pay to have it done for you. If you are relatively good with technology or have someone who can help, you can purchase or find old hardware that is compatible with what you need. The technology you choose must also allow you to pull files to a more modern location. So, for example, if you have Zip disks and no way to use them, you could purchase or borrow a Zip drive and connect it to a computer that also has a USB port for flash drive access. In this way you can move things from the Zip drive to the flash drive and then to your current computer. Chapter 8 has additional tips for transferring legacy media to your computer.

Maintenance

As part of your plan, you will want to decide how much (or how little) maintenance is needed to keep things running and organized the way that you want it. If there are any parts of your backup system that are not automatic, then schedule a time to make those updates. In your written backup plan, list all of the things that you will not be backing up, those that will be automatically backed up, and those that require you to take extra steps. As things change, make changes to your plan so that you remember how things work. For example, you should update the plan if you change phones or stop using a device. More maintenance activities may be added to your list as you look at the more complex locations covered in later chapters.

TIP: Setting aside a time to do maintenance along with another activity can help. For example, if you have an older mobile phone and no automatic backup is available, set aside time to back up your phone's photos once a month when you pay your rent. Linking any needed activities with something you are unlikely to forget will help ensure that the work gets done.

Implications for Your Physical Stuff

Having a backup system in place for digital files should help you manage your physical files as well. Now that a backup is in place, any time or effort spent digitizing physical items will automatically be protected. Since there is nothing more demoralizing than losing a large batch of old photographs that it took weeks or even months to digitize, having a backup system can encourage people to actively work on any physical items they want to convert to digital files. There is less risk when things are backed up twice. Once you have full faith in the new backup system, most people feel comfortable getting rid of their paper files, depending on what they are. Obviously some items, like legal documents, may be candidates for digitization as a backup, while the paper copy is still retained. More on documents and files will be covered in chapter 2.

TIP: Once you have a double-backup system in place, you should feel confident in getting rid of old backups on CDs, flash drives, or other media—but that is really up to you. If you feel more comfortable holding onto it, mark the media in some way so that you know it has been saved elsewhere and you don't need to duplicate your efforts later.

What about the Unimportant Stuff?

For items that ranked low on your priority list, you can choose how best to deal with them depending on how much time and computer space you have. For devices to run efficiently, it isn't ideal to wait until you are nearly out of space to weed files. For purposes of space, you will want to delete unwanted files regularly. For example, if text messages aren't worth keeping, delete them in batch every once in a while or, if you prefer things tidy, consider setting aside time to delete things on a weekly basis. You may also find that you don't want to commit to the maintenance activities involved in backing up some items, such as voicemail or photos on an old phone. If you are not prepared to upgrade or purchase applications to make the process easier, then you may decide to change what you back up. That is OK, too, since personal digital archiving is all about you making your own choices for what to keep and protect.

TIP: Get into the habit of cleaning up your phone while you wait in line at stores. Get rid of fuzzy photos, unwanted texts, and junk mail during these small chunks of time. Even better, do it while waiting for your flight at the airport.

WHAT YOU SHOULD HAVE ACCOMPLISHED AFTER READING THIS CHAPTER

After following the steps in this chapter, you should:

- Know what digital items you have, and where they are.
- Have a double-backup system in place to protect your items.
- Have a plan for maintaining your system.
- Have scheduled times for periodic maintenance, if any.
- Know which things aren't important, and can be ignored or deleted.

Having backups in place and a plan for getting things moved to a centralized location makes it easier to tackle other issues, like organization. That's what happens next. Since things are protected, it is a good time to get started on cleaning things up. Chapter 2 will address file naming, organization, and structure so that you can find and understand the things you have backed up and protected.

NOTES

1. Catherine C. Marshall, "How People Manage Personal Information Over a Lifetime," *Personal Information Management*, edited by William Jones and Jaime Teevan (Seattle: University of Washington Press, 2007), 57–75.

2. "Find or Recover a File," Google, accessed May 20, 2016, https://support.google.com/drive/answer/1716222?hl=en.

3. Richard Rinehart and Jon Ippolito, *Re-Collection: Art, New Media, and Social Memory* (Cambridge, MA: MIT Press, 2014).

4. "Computer Backup System Buying Guide," *Consumer Reports*, March 2015, http://www.consumerreports.org/cro/computer-backup-systems/buying-guide.htm.

5. Brandon Butler, "Cloud's Worst-Case Scenario: What to Do If Your Provider Goes Belly Up," NetworkWorld, June 2, 2014, http://www.networkworld.com/article/2173255/cloud-computing/cloud-s-worst-case-scenario-what-to-do-if-your-provider-goes-belly-up.html.

6. Thomas McMullan, "How to Access iCloud on a PC," Alphr, January 14, 2016, http://www.alphr.com/features/390340/how-to-access-icloud-on-a-pc.

7. "What Can 49,056 Hard Drives Tell Us? Hard Drive Reliability Stats for Q3 2015," Backblaze, October 14, 2015, https://www.backblaze.com/blog/hard-drive-reliability-q3-2015/.

8. "Computer Backup System Buying Guide," *Consumer Reports*.

9. Ibid.

10. Ibid.

2

Naming, Structuring, and Organizing Files

Next to backing up files to prevent loss, naming and organizing files is the most important thing that you can do to ensure long-term access to your digital items. Deciding what files should be called helps you to recognize them, search for them, and have them appear in a list the way that you want them. In fact, depending on how you name your files and how you use your computer, you may find that you do not need extensive folder organization: file naming on its own can be enough for some people. That will depend on your personal needs, like so many other parts of personal digital archiving. It may also depend on what other metadata your files have. Don't worry: this chapter also explains metadata, and why it matters when finding and organizing your stuff.

There are a number of organizing and naming options, but ultimately it will depend on you to choose the best way to name and manage your digital files. Like naming your children, there are personal reasons why you might choose something different than your neighbor. For one thing, names have different meanings to different people. Some people might keep their veterinary records in a folder called "Pets" and may have a different folder for "Animals" because some people make a distinction between the two, while others do not. Some people may want their photos to easily organize by date in each folder, while other people like locations or descriptive names. It is up to you, although there are some tips and ground rules for coming up with a system that works consistently. Naming conventions and organization can apply to many areas of life, including your home and work life, your physical filing system, and your email accounts. Each of these topics is covered in this chapter.

FILE NAMING

Being able to find and recognize your files is crucial. Rinehart and Ippolito suggests that, "If you can't find it, it's not preserved."[1] Finding, and also recognizing, the file you need, when you need it, is why you should give some thought to file naming. Many of the worst file-naming issues are not the fault of the user (at least not directly). In many cases, the long, unintelligible number-letter strings that some of our files have were assigned by a website or program. Sometimes we get files from others and these may have names that are OK—but they are not what we would have named them. We download and save these files and do not take the extra time to change them. Another frequent issue is consistency: we don't have a plan for how to name our files, so we do whatever comes to mind at the time. The lack of consistency plus multiple files coming from outside sources results in a confusing collection of file names that do not work well together.

Look at figure 2.1 to see many of the common issues related to file naming. You can see many examples of files that do not have descriptive names: you simply cannot tell what they are by their name. This forces you to open them to identify them, which can be very time-consuming. Other files in this example have been assigned

Figure 2.1. View of an unorganized folder with poorly named files

names automatically by a computer. You can easily find those because they are long strings of characters and are not like anything a human being would normally name something. This is common for image files created by digital cameras and phones, although software and phone applications are improving. Finally, the example also lists a few things that are somewhat descriptive, but are related to other files that have not been named using the same system. This inconsistency means that users must look for a file in multiple different ways, and this also wastes time.

Parts of a File: Name versus Extension

Before we talk further about what to name files, it is important to understand at least some basic information about file extensions. A file extension is a short code after the period of your file name that tells your computer what sort of file it is. For example, a file might be called JimSmith_Resume.pdf or 2016Budget.docx. In these files, pdf and docx are the file extensions and should not be changed if you change the rest of the file name. Depending on how you change the name, your computer will warn you of the risk to avoid accidentally changing the extension. If you have ever tried to open a file and had your computer show an error message that says "File type not recognized" or similar, that is because the file extension is not a common type and you may not have a program that can open it. It is important to know that computers rely on the file extension to know what kind of file you are using, so *file extensions should not be changed in most cases*. It is also helpful to know that you can keep multiple files of the same name in a folder if they all have different file extensions. This may come in handy when you want to create the same file in two different formats for sharing. For example, you might have a word processing document to edit your resume, which you can also save as a harder-to-edit PDF file to share. Where two files are the same type, however, your computer will not allow you to keep two files of the same name.

Other than the name extension, you can change the name of a file in a number of ways:

1. On a Windows PC you can right-click on a file and choose Rename.
2. You can choose Properties from the right-click window, which allows you to change the name from the General tab.
3. You can single-click on the current title of the file either on your desktop or in a list of files to highlight it, and then single-click the file again to rename.

Directions for batch-changing groups of files are provided later in the chapter.

Basic Rules for File Naming

- Be descriptive. A file named Recipe.pdf is not nearly as helpful as a file named MomsVeganTiramisu.pdf. Avoid acronyms unless you are sure that you will remember them or they are easily recognizable. If other people use your computer or may need to find files for you, make sure your names don't have special meanings that only you can recognize. Ask yourself: "If I see this file name in 10 years, will I know what it is without opening it?"

- Be consistent. If you have 400 photos in a folder and half of them were named with the photo location and the other half were named by the photo date, then you will have difficulty finding the photo without looking in two places.

- Avoid special characters. The following characters should not be used in file names because they can cause errors when they are uploaded, downloaded, emailed, or converted: slash [/]; backslash [\]; question mark [?]; percentage sign [%]; pipe [|]; asterisk [*]; quotation marks [""]; colon [:]; less than [<]; greater than [>]; and period [.]. These characters have specific functions in many systems and create errors when they are used in general file names. Some programs will not allow you to use them, but others will. They may not cause trouble right away, but it is best to avoid them so that you do not have a problem later.

- Use underscore rather than spaces. This is especially true if you upload files to the web and social media often. Spaces sometimes get converted to strings or characters when they are saved on the web. This can result in unnecessarily complicated web path names. Consider using an underscore [_] or dash [-], or running words together without spaces. Search will still find individual words that are run together in this way. You can also use selective capitalization where words are run together with each first letter capitalized: for example, SmithResume.doc. This helps the human eye find the words even though they are run together, and can be helpful when browsing files. Be aware, however, there are some computer systems and programs that will change inconsistent file names to all capital or all lowercase.

- Avoid names over 21 characters; they can be problematic. As with many unique characters, this may not cause trouble right away, but can cause issues in the future when programs you need to use have character limits. Some systems will cut off file names at 21 characters. If you want to use long names in order to make something searchable by including some words, consider using different file metadata options instead. Some options are described later in this chapter.

- In general, use letters, numerals 0 through 9, and – or _ for most file names, of all formats.[2]

Naming for Browsing

When deciding how to name certain types of files, you will want to consider how you will use the files in question. If you are likely to open a folder and browse the items—which is common in cases of recipes or saved articles for reading later—then you will want the names to be browsing-friendly. If you frequently browse your files rather than searching for something specific, emphasis should be placed on descriptive naming.

Naming for Searching

If you do not have time to organize files and regularly rely on a search to find your items, limited organization may allow you to still find what you need using the search function alone. Russell and Lawrence note that our terrible human memories make search a great option in many cases: "Personal information management tools can rescue human memory from the transient mistakes and failures that are commonplace, giving a remarkable ability to look up things quickly."[3] Your computer's search box (both Mac and PC) will look at the file name, tags, author, type of file/file extension, and readable text inside your documents. That means that a file name on its own does not have to have everything you might search for to find your document. Keep that in mind when naming a file: it should be descriptive, but doesn't have to have everything. Be aware of your default search options, and the built-in search functionality on your computer. Figures 2.2 and 2.3 show some of the search options available using Windows PC and Apple built-in search.

Naming for Sorting

When you look at a list of files on your computer, you have the option of seeing them in a list that contains additional helpful information. On a Windows PC, you can select the Detail option to see the list of file names, date last modified, type, and size. A Mac has similar options. You can sort by any column by clicking on the

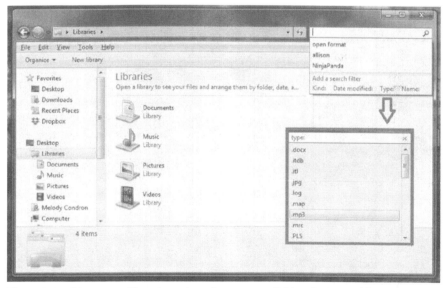

Figure 2.2. Numerous filters/search options are available when searching a Windows PC.

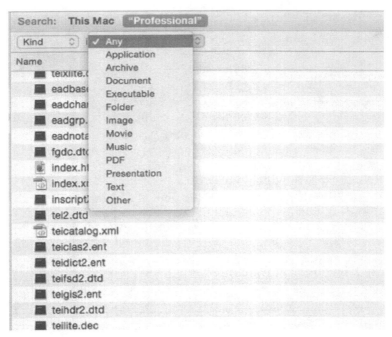

Figure 2.3.   Apple drop-down search filters also offer ways to find and limit file search results.

header/title of the column. That means that you can sort by date modified to find files that you worked with recently, or click it twice to see files in reverse order (oldest at the top). Similarly, you can sort by file type to find Word documents, or by size to find your largest or smallest items (this can be helpful to find space-hogging design files). The most useful sort, however, is sort by file name, because this allows you to take advantage of naming conventions that you have put in place. Files will sort alphabetically and numerically, which means that you can name files in such a way as to make them all sort together. For example, if you want all of your budget files to come up together, you can add the prefix "Budget" to the front of each file. Budget_2016, Budget_2015, and Budget_2014 would all sort next to each other in a much larger list.

Consider numerical sorting when adding dates to your file names. Using a structure such as 12-15-2015 for dates will mean that your files sort by month: all of the December files (12) will be next to one another when you sort by title. If you want files to sort in date order, you will need to use a Year-Month-Day order. For this system to work, you must use numerical dates because spelling out months will result in months being sorted alphabetically. You must also use a zero [0] in front of single-digit months and days. Avoid two-digital number years, as they can be confused for days, months, or non-date related numbers. The YYYY-MM-DD format (with or without dashes) works for many people and is a common way to sort photos or other large numbers of files collected over a period where date may be a distinctive element.

Combined with the search function, sorting can be very helpful. Many computer users can find what they need by searching and then sorting. If this system works best for you, then extensive file folder organization may not be as helpful for you as for some other computer users.

Version Control

When working with others or alone, sometimes you will create many drafts of a document before it is finished. Many people like to keep several drafts so that they can go back to an earlier version if they change their mind. In other cases, someone makes a copy of a document and adds his or her own notes or corrections. In all of these cases, it is important to consistently keep track of which file is the newest and which version it is. This is called version control.

Here are some tips for improving version control and tracking, so that no one loses work and the latest file is always easy to find:

- Number the current version even if it is the first draft. A name like NASPA_v1 can work, or you can use the date; for example, NASPA_20160101.
- Get into the habit of saving the file as a new version as soon as you open it.
- For groups: note that there is no great way for multiple people to make copies of a document, make edits, and then merge all of the copies. If this is your intention, consider using a file collaboration tool like Google Docs or Basecamp.
- If working with a group, discuss file names and version numbers before your project begins. For example, if Kevin Jackson adds to the document and sends it back, he could return the document with his initials at the end: NASPA_20160101kj.
- For individuals: note that making multiple drafts may be more complicated than necessary. If you use a backup system like Apple Time Machine, multiple drafts of your document going back for weeks will be automatically saved for you. Consider whether you really need to save different versions. Version control can become confusing and is often the cause of confusion when updates are made to multiple versions.

TIP: consider writing down your file-naming procedure with examples, or emailing the procedure and examples to the group if you are working with a team.

Naming and Metadata

Metadata is the data about data: it includes the details about your file and the information it contains. According to Rinehart and Ippolito, "Another way to put it is that metadata provides context . . . metadata reveals a level of meaning that is implicit in the data itself and easily understood by humans but not by computers."[4] Naming is just one form of metadata, and knowing more about metadata can help you make better naming choices. The name of a file is a searchable form of metadata, but it isn't the only one. Search will also find text within a text-based document, and tags that have been added to an item manually. Knowing what search will look at helps you to make educated decisions about file naming. For example, if you have written a paper with multiple references to Jane Goodall but she is not the main subject of the paper, you need not include her in the title: a basic search for Goodall will find her name inside the text document.

Tagging is an additional way to reduce the need for overly long file names. Figures 2.4 and 2.5 show the tagging fields on a Windows PC and an Apple computer, respectively. By adding tags for subjects and topics you may want to find later, you can create better search results. This is especially helpful in cases where the text within the file and the file name may not have the information automatically. For example, giving recipes tags such as Lunch, Dinner, or Dessert would allow you to identify recipes of a certain kind, though that information might not be in the recipe itself. Tagging allows you to add that information to the file without using up the limited characters recommended in title length. Some research has suggested that effective

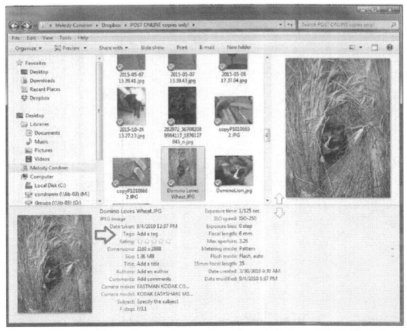

Figure 2.4.   To add tags/keywords to a file using a Windows PC, right-click on the file and select Properties.

Figure 2.5. To add tags/keywords to a file using an Apple computer, open the document and select Control + I to view the File Inspector.

tagging and file naming can actually be more time efficient than file folder organization systems.[5] Beyond built-in tagging, there are many other options for embedding metadata in files. Additional details about tagging and photograph metadata will be covered in chapter 5.

HOW TO FIX IT

So how do you fix things now, when your computer is already full of poorly named files? There are a number of things that you can do to start:

1. Decide what naming scheme or rules work best for you, and stick with it. Remember, you might have different rules for some things (like photos).
2. Name all new/incoming files using your new naming scheme.
3. Get rid of files you don't need rather than waste time renaming them. Alternately, you can just rename your important files and delete unimportant files as you come across them.
4. Batch-rename files where possible (covered in the next section).

Fixing Files in Batch

Renaming files one by one is a daunting task. Luckily, most computers come with built-in programs to assist you. If the built-in programs are not as full-featured as you need, then many programs for purchase are also available. For purposes of renaming files in groups it is helpful to have them all in the same folder without any additional files, in order to avoid accidental name changes. Windows and Apple computers each have different functionality, with Apple offering more options and control. Figure 2.6 shows an example of the Windows rename files function changing a group of poorly named photographs so that they all have the same title of 2015_HI_Vacation with numbers to indicate each different file. The rename function is easy to use and find by selecting the files to change and right-clicking to see the Rename option. On an Apple, you also select the files you want to change and right-click, then choose Rename ## Items. In figure 2.7 you can see that Apple allows for some additional options, including appending or replacing text with selected text. This is especially helpful when you do not want to delete existing file names, but would like to add just a date or note to the front of the file name for searching or organizational purposes. Again, if the built-in functionality on your computer does not meet your needs, then many low-cost options exist.

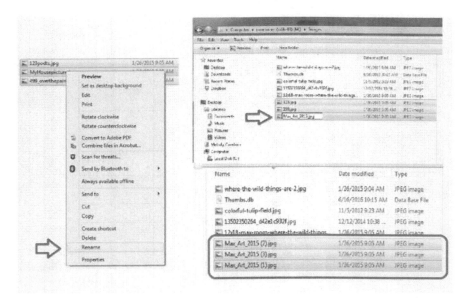

Figure 2.6.   Options for batch name changes on a Windows PC

Figure 2.7. **Options for batch name changes on an Apple computer**

FOLDER STRUCTURE AND HIERARCHY

It is impossible to have no file structure on your personal computer, since some structure is built in. Like a physical filing system, creating folders with clear names can help you browse and find the files you need. However, just like with paper files, this system only works when items are all put into the folders and the folders make sense. This is why so many people are frustrated with digital folder systems and resort to using the search function rather than addressing the folder system itself. If you find the upkeep of an extensive folder system daunting or confusing, then consider focusing on better naming and tagging for your documents. This is a legitimate approach: many information researchers argue that computer folders' basis in physical documents is a limitation, and that digital use of these folders will die out in preference of tagging.[6] Whether the big folder die-out happens or not, many computer users find folder structure straightforward and familiar. You may choose to use them diligently, or you may use very few for special projects: the choice is up to you. If you do choose to avoid folders for the most part, you might consider creating some very broad folders at a minimum to keep items from cluttering the Desktop and your Documents folder. Broad categories like Current Projects, WORK, and HOME work for some people.

If you want to make the most of your organization system, enjoy browsing, and want things to be "put away" on your computer, then you will need to develop a folder hierarchy. Chances are, you already have one (though in many cases you have not kept up with it). Most digital-file folder systems have developed over time and can appear haphazard after the fact. If you want the folder system to make more sense, you will need to review your folder names and rename them. The beauty of digital folders is that you can rename them often and easily. If you find that you are reluctant to put items into the folder they should probably go into, then perhaps the name is not accurate for your purposes.

TIP: You can use the same folder names for your digital files as for physical files, if that makes sense to you. But you don't have to!

Here are some basic suggestions for folder creation:

- Top folders should be broad and easy to understand, and folders nested inside should get more specific as they go.

- Consider alphabetical sorting when creating folders, just as you did with file names.
- Ideally, for both visual clarity and organization, all files should be in a folder. Only files that are in use at the moment (e.g., just saved or created) should be "loose" on the Desktop or in your Documents folder.
- If you have a file that does not logically fit into your existing folder system, create a new folder rather than overthinking where it should go in your existing system—even if that is the only item inside so far. Often you will get more items that will fit there later, or you can always change the folder name later if you need to.

TIP: If you are overwhelmed by the clutter of documents already on your computer, consider creating some temporary files marked To Sort. You may even have multiple files like this, with a Photos to Sort file, Work to Sort, and so on. While this is not a long-term solution, it does help many people feel less overwhelmed with the overabundance of files that sometimes clutter their Documents folder and Desktop.

Common Practices

The two most common practices in folder organization systems are subject and file type organization. Both scenarios can be used together to some extent. Explanations and examples of each are provided. Since each user will have different files and needs, it will be up to you to decide if either method or both methods should be part of your system.

Scenario 1: Folders by Subject

Organizing files by subject is an intuitive method, and is the basis of most filing systems (physical as well as digital). Ordinarily this type of organizational system would have folders with broad topics, with additional, more specific folders nested within. For example, a folder called My Writing could have folders inside for Poetry, Short Fiction, and Magazine Articles. This works for most documents, but get tricky when working with images and multimedia. The subject of something is not the same as what something *is*, which can be confusing. For example, a folder called Photography could contain resources about photography or photographs. Figure 2.8 shows a subject-focused folder organization system.

Scenario 2: Folders by Type of File

Organizing by broad files type is not as common, but is sometimes part of an overall system. For example, folders labeled Photos, Audio, Videos, and e-Books may be helpful to some users and could be added to a mainly subject-based organization system. Since these tend to be very broad designations, and users tend to have many

Figure 2.8. A subject-based folder organization system

files in each, more specific top-level folders may be in order. For example, a Documents folder might be split into Work Documents and Personal Documents instead of having an extra layer of folders. Figure 2.9 shows an example of a folder organization system that is organized by type of file.

EMAIL ORGANIZATION

Organizing email accounts can be similar to organizing files on your computer, with a few caveats. Email folders and organization definitely have a few differences that can make it much more complicated or much easier to organize correspondence, depending on your point of view. These include automatic email organization rules, lack of file name control, and the trend toward tagging and moving away from folder use.

A great benefit of email over regular computer files is the ability to set up rules for incoming mail. All email systems have functionality that allows for emails with certain subjects or from certain addresses to be sorted before they are even read. If set up properly, this type of rule process can greatly improve organization with a limited

Folders	Folders	Images
Audio Books ▶	Autobiogra...min Franklin ▶	Autobiograp...Franklin.jpg
eBooks ▶	Malice Dom...tery Stories ▶	**Music**
Home Docs ▶	More Malice Domestic ▶	Autobiogra...-Part01.mp3
Home Videos ▶	Mystery Sto...olet Strange ▶	Autobiogra...-Part02.mp3
Movies ▶		Autobiogra...-Part03.mp3
Music ▶		Autobiogra...-Part04.mp3
Photos ▶		Autobiogra...-Part05.mp3
Work Docs ▶		Autobiogra...-Part06.mp3

Figure 2.9. A type-based folder organization system

amount of time spent on setup. For example, if all emails from your closest friends and family were sorted into a high-priority Read Me First folder or tag, this would allow you to skip wading through other emails to get to things you value.

One of the difficulties in organizing email messages is the lack of control: email is a cooperative effort between you and everyone you communicate with online. Without any control over the name or subject line of incoming emails, search is often more difficult. Being aware of how you yourself search your email, you can consciously decide to use better subject lines in email communication. However, searching email messages will never be perfect because most people do not follow any guidelines when emailing.

Finally, in recent years, services including Gmail have begun moving toward a tag-based system of email organization. While you can still create folders, emails can be tagged to belong in multiple places. While this is good if you do not remember which tags you used, there is no longer a firm feeling of "location" as tagging takes over. It will likely be years (or never) before folders disappear from email services, but this is a growing trend in email management. Becoming familiar with tagging and recognizing its usefulness in relation to file management will help in any possible transition. Thankfully, email providers usually have robust and effective search tools to assist with email recall.

WHAT YOU SHOULD TAKE AWAY FROM THIS CHAPTER

After reading this chapter you should have:

- Learned tips and things to avoid when naming files.
- Discovered how tagging and file naming can complement one another for better search results.
- Learned how to rename files in batch.
- Considered whether extensive folder organization is more effective for your purposes than relying on search and sort functions.

NOTES

1. Richard Rinehart and Jon Ippolito, *Re-Collection: Art, New Media, and Social Memory* (Cambridge, MA: MIT Press, 2014).

2. Mike Casey and Bruce Gordon, *Sound Directions: Best Practices for Audio Preservation*, 2007, http://www.dlib.indiana.edu/projects/sounddirections/papersPresent/index.shtml.

3. Daniel M. Russell and Steve Lawrence, "Search Everything," *Personal Information Management*, edited by William Jones and Jaime Teevan (Seattle: University of Washington Press, 2007), 57–75.

4. Rinehart and Ippolito, *Re-Collection*, 57.

5. Andrea Civan, William Jones, Predrag Klasnja, and Harry Bruce, "Better to Organize Personal Information by Folders or by Tags? The Devil Is in the Details," *Proceedings of the American Society for Information Science and Technology* 45.1 (2009): 1–13.

6. William Jones, "How People Keep and Organize Personal Information," *Personal Information Management*, edited by William Jones and Jaime Teevan. (Seattle: University of Washington Press, 2007), 35–56.

3

Legal, Financial, and Medical Documents

It is not uncommon to have a mix of digital and physical items (or just physical items) when it comes to legal, financial, and medical documents. This may be because so many of the most important documents—such as wills, medical records, and birth certificates—come in physical form and are rarely issued digitally. In fact, some of these documents are invalid when digitized: so why would we scan them? Concerns about privacy and risk of identity theft may also lead some people to keep these items undigitized. What this often means is that we only have one copy of these important documents that can be difficult or impossible to replace, and that puts us at great risk of loss in the case of emergency.

The Federal Emergency Management Agency (FEMA) emergency preparedness guide recommends keeping insurance documents and vital records in a safe place, such as a safe-deposit box away from your home.[1] FEMA also recommends watertight containers for these documents, and adding photocopies of these documents to your emergency kit. While this may work for some people, especially those with only a few documents to worry about, it is a more difficult task for others. Keeping current financial documents in a safe-deposit box and making physical copies becomes arduous when your situation changes regularly: you must remember to update copies and documents, and that often does not happen consistently. There may be nothing worse than being the victim of an emergency, only to find that the insurance or documents you need to receive assistance are not in order. Insurance companies and other emergency preparedness recommendations include keeping documents somewhere handy to grab as necessary in case of flood, fire, or other calamity. However, this, too, does not help if you are not home when a disaster strikes.[2] It is true that keeping digital copies cannot substitute for all physical documents, but it can help, and is worth considering as part of an emergency plan. It may also be possible to digitize some documents and not keep the physical copy, which can simplify your life by reducing the clutter of paper files.

COPIES CAN HELP

Though copies and scans cannot always take the place of originals, copies can often help you recover from emergency if your original is lost. The U.S. State Department, for example, recommends that you make color scans of your passport when traveling internationally so that the numbers and information on the document can be looked up or used for verification if your original passport is stolen while you are abroad.[3] The same is true of many other documents: the originals (and copies) have codes and information that we do not have memorized and may assist issuing bodies when these must be replaced.

Keeping a digital copy of your important documents also allows them to become part of your automated backup system (see chapter 1 to find out more about setting up an automated backup system). Unlike creating photocopies to keep in two different locations, this means that any updates to a document are automatically updated in your two backup locations. This will not work for passports and marriage licenses that do not change over time, but it will work for financial and many other documents. For example, if you receive physical letters in the mail from your insurance company outlining recent costs and reimbursement, scanning that letter will allow you to have a record of those transactions.

TIP: If you do not have a scanner, tablets and phones now have many applications available that do a decent job of creating a scanned image. Look for apps that save to PDF and have Optical Character Recognition (OCR). OCR will allow the program to "read" and recognize printed text rather than saving the document as an image. While having a photo of a document is better than nothing, having text saved in the scan will allow for better searching.

For financial statements that come regularly (bank, credit card, and bills), keeping digital copies is a much better option for many people. These documents are usually online at your service provider's website. However, some institutions only keep a certain number of years available, and they may become inaccessible if you switch providers. If statements are available as PDFs, you may prefer to download them on a scheduled basis so that you have them if you switch providers or access/website offerings change (more on this later in the chapter). Keep in mind that not all individuals need to keep all of their statements indefinitely. If you are self-employed or have complicated financial dealings, it may be to your benefit to keep documents in case you have questions later.

TIP: Consider downloading last year's records when you do the year's taxes. Since that is often a time when you must review finances and accounts to report interest income, making a habit of downloading statements can help you save and keep these documents annually rather than trying not to forget what month you last downloaded throughout the year.

Remember that determining which documents are important and should have a digital copy is a personal choice based on your situation. This chapter offers recommendations, but these may not fit with your individual needs. The recommendations are for a general audience and include best practices that are generally accepted. In most cases, these recommendations are the minimums: they say how long to keep these items if no other special situation applies. Certain types of documents should be kept for a specific period of time, usually based on tax and legal requirements. In other words, you may need them within this time frame if you are audited, or need to prove something for legal reasons. Table 3.1 illustrates suggested retention times for different kinds of documents. It also notes whether items should or can be digital or physical. Sections later in this chapter will cover specific document types separately to explain some of the issues specific to each, and will briefly discuss what options are available for digitizing your documents and either encrypting or password-protecting them for safety. Finally, this chapter will explain the idea of legacy planning for your digital documents. Legacy planning will review your options for allowing others access to your digital items in case of emergency or death.

LEGAL DOCUMENTS

FEMA lists most of the following legal documents when discussing what to protect for emergency planning and preparedness:[4]

- Deeds
- Wills (might also include living wills, power of attorney, or advance directives, if applicable)
- Property records
- Passports
- Social Security cards
- Driver's license or state identification card
- Birth certificate
- Marriage certificate
- Divorce decrees
- Adoption paperwork
- Military discharge

Most legal documents are only valid in their original physical form. In some cases this is because of special seals, watermarks, signatures, or seals that are added specifically to avoid document fraud. This is a good thing: you don't really want extra copies of your passport all over the place. However, since these items are sometimes one of a kind, it can take a lot of effort to get a new one if they are lost or destroyed. Most should also be kept indefinitely, so losing them usually means you will need to get a replacement.[5] Despite not taking the place of an original, keeping digital copies of legal documents protects the user and can make it easier to replace the document

Table 3.1. Document Retention Guide

Document Type	Period of Retention	Keep Physical Copy if Digitized?	Concerns/Directions
Financial			
Bank statements	1 year to permanently	No	Statements over a year old with no major or tax-related purchases can be shredded.
ATM receipts	Until reconciled with statement	No	
Credit-card statements	7 years	No	If you are sure that nothing purchased with your credit card is tax-related, then statements can be retained for 2–3 years.
Retirement account	Until retirement or account closes	No	Keep quarterly statements until reconciled with annual, then keep annual only.
General bills & receipts	Keep for large purchases as long as item is retained	No	Large-purchase receipts/bills should be kept with insurance papers to prove value in case of loss.
Taxes & related documentation	7 years	No	The IRS has 6 years to challenge your return for underreporting, more if you knowingly filed a fraudulent return.
Utility bills	Keep during discrepancies or until payments are posted	No	
Vehicle service payments	As long as you own the vehicle	Yes	These may be needed for warranty or for next owner, if sold.

Legal

Mortgage & loan documents	Permanently	Yes	Keep the signed loan paperwork even if you have a digital copy.
Car title	Permanently	Yes	
Identification documents	Until replaced	Yes	Some states have laws that require you to destroy old identification cards when a new one is received.
Insurance documents	Until claims period after policy has ended	Yes & No	Keep all signed documents. Other documentation can be digital. Even if you cancel a policy, there is a statute of limitations for late claims. Check your policy.
Wills & estate documents	Permanently	Yes	
Vital documents (birth, marriage, immigration, divorce, & death certificates)	Permanently	Yes	

Medical

Insurance documentation, including premium statements	5 years	No	If these are the only records you have of medical procedures, medications, or other medical services, consider retaining or otherwise note dates.
Medical expense documentation	7 Years	No	Keep 7 years for tax purposes.

if it is lost or destroyed. In some cases, copies may work as well. For example, many utility and other companies will accept a digital scan or photocopy of a marriage certificate as proof of name change for accounts. Also, in cases of large-scale emergency (such as community flooding or tornado), some rules may be relaxed; certainly a duplicate of the original is better than having no documentation at all.[6] Consider the temporary printed driver's license that you get before your actual license arrives in the mail: this document will protect you if you are pulled over, because someone can check your credentials based on the information provided on the printed copy. Despite the usefulness of copies, many people never create duplicates of their important documents and only have the physical, original item.

If you create digital copies of your legal documents, make sure that your backup systems are local (flash drive or other local non-cloud backup) or encrypted to avoid identity theft. Encryption converts documents into ciphertext while locked, and people cannot read the documents without the password key, protecting your private documents while they are stored. Many cloud backup systems have minimal encryption, so you will need to check with your provider. Most paid cloud services offer some encryption services. Encrypting individual Microsoft documents is explained in detail later in this chapter.

In the case of the legal documents listed above, none can be completely replaced by a digital copy. In other words, you cannot scan these documents and then dispose of the original document. Creating digital copies of legal documents allows you to share them, improve your own access, or to protect yourself in case of emergency. Sharing may be advantageous in the case of wills, parental agreements, or other documents that involve multiple people. While the actual signed, legal document is the item that has legal power, an individual might share a copy of his or her will with many family members as part of their estate planning. The signed physical document might be kept with a lawyer, but all family members would be familiar with the will because the copy was shared digitally. As for access, it may be easier to find and review your will and other documents if you have a digital copy. Finally, in case of emergency, having a digital document copy that is part of your backup systems means you cannot lose it—even if the whole town, including your lawyer's office and your bank's safe-deposit boxes—are under water. Since no one is immune to natural disaster, protecting your most important documents may be something you want to do, like buying insurance.

FINANCIAL DOCUMENTS

Financial documents are different from legal documents in a number of ways. In many cases financial documents are records of accounts and transactions. They are mainly informational. Documents like bank and credit-card statements can usually be reissued if they are recent. Some of them can be digitized without keeping a paper copy, which means that you can clean up piles of older paper statements if you would rather retain this information on your computer than in paper files. Almost all financial institutions now allow you to receive digital statements. If you still prefer to get a paper copy but would like a digital backup, you should be able to download statements from your provider's website rather than scanning in your paper copies.

TIP: Downloaded statements are better than scans of paper documents, since digitally created documents usually have searchable text: that means you can search for the words in the document when you save them on your computer.

FEMA lists the following financial documents as important to protect in its emergency preparedness documentation:[7]

- Insurance policies
- Household inventory list, photos, or video (to support insurance claims)
- Bank and credit-card account numbers
- Stocks and bonds

Other financial documents that are important but perhaps not critical include:

- Bank statements
- Bills and billing statements
- Receipts for large purchases
- Retirement account numbers and statements
- Taxes and related documents
- Loan documents (also considered legal documents)

Insurance and Related Documentation

Most adults have some form of insurance, such as automobile, homeowners/renters, and life insurance. While you may feel like you can get a copy of your policy at any time and may not need to have a copy on hand, consider that emergencies are the worst time to have to wait for an insurance company to find and send you your policy if it isn't immediately available. You may need it more quickly than your insurer can provide it, and the company may also be slower to provide service if you were part of a larger natural disaster because it will be working with a lot of people at once. Also consider that you may avoid unnecessary calls and questions by knowing the direct phone number and policy number for your insurance. Finally, make sure that your family members know which insurance company you use so that they can call for information if you are incapacitated during an emergency. Nolo, a publisher of legal guides, reports that millions of dollars of life insurance money is never claimed because beneficiaries do not know about the policy.[8] If you are tempted to get rid of the paper copy of your policy, make sure that your loved ones have the information they need.

You may have online access to your insurance information, including the full policy, through your account login. However, you will also receive a physical confirmation in the mail with the policy documents. If you would like to have a digital copy of your policy, contact your insuring agent to see if they can email you a PDF version, or you may be able to download it from the website. It may prevent a lot of hassle to save the document where you know you will have access. Still, easily obtaining a copy of your policy applies to most insurance policies including automobile, home or rental, and life insurance.

TIP: It is unlikely that you will be able to get a comprehensive document concerning your health insurance. Laws and coverage related to health insurance change often. The closest thing that you will receive from your employer or provider is likely to be an annual guide to your health insurance coverage; sometimes this guide is available online or by request from your employer's human resources department. At the very least, having a copy of your insurance card will help you know which policy you hold, and with what company.

Bank and Credit-Card Documentation

Banking and credit-card documentation are increasingly digital, and you may be able to keep most of your account documents on the computer. This is often more complicated that it sounds. If you currently receive online statements from a financial company like a bank, you may notice that they send you an email saying that your account statement is available. However, these email messages do *not* contain the statement. You must log in to retrieve your statement and view your account. The email is just a reminder, and does not need to be retained as it contains no information. While this practice is meant to protect users from accidentally having banking information exposed in an email account, it often means that people do not view or keep their account information once they switch to online statements. This can be problematic, because some banks (most often smaller regional banks or credit unions) only make statements or transactions available for a few years. This may not be an issue for people who never see the need to look into old statements, but others like to have the record of account transactions for longer periods. You might need older statements for a variety of reasons, including tax audits, financial planning, or proof of ownership or payment claims.

Documents related to loans and mortgages are different than other banking documents, because these count as legal documents and normally need to be kept in physical form, although much of the loan and financial details will be available online as well. Loan documents should be retained until loans are completely paid.[9] Your opening account documents are provided in paper form because they are also legal documents showing that you own the account, which is why these require a signature. However, many people do not retain these documents because the banks keep copies and the documents are rarely used.

TIP: Bank statements are not the same as logging in to an online account and seeing your balance. Statements are issued monthly for most accounts and are usually available as PDF downloads. Unlike the online login system, they show a point-in-time "snapshot" of your account rather than the ever-changing account view posted online. Some transactions (like mortgage applications) will require bank statements.

Investments: Retirement Accounts and Portfolios

Like bank statements, retirement account and investment statements can usually be maintained digitally. Statement alerts may be mailed physically if you have not opted in to digital statements. If you do receive digital statements, you likely receive email alerts that your statement is available through web account login. Also like bank statements, there is a difference between the overall account view and the actual bank statement. Periodically downloading statements is recommended in order to track how your investments are allocated. It may also be helpful to download investment prospectuses from investment websites if you are involved in selecting your own investments.

Taxes and Tax-Related Documentation

The Internal Revenue Service (IRS) offers guidance on what kind of records should be kept for tax purposes, and for how long.[10] The IRS recommends keeping tax returns for at least three years.[11] Supporting documents should also be kept during that time period, including receipts for deductions, worksheets, proof of payments, and even (as of 2016) proof of insurance coverage. If you do your taxes online or use a service, make sure that you request or download a copy of your taxes, preferably in the format of the tax forms themselves. Requesting copies of your older tax forms from the IRS is not an easy process, and if you are being audited, it is much more helpful to have your own copy to look at and prepare. Beyond the tax forms, anything listed on the forms (from interest payments to income) should be verifiable later. Keeping the records, receipts, and statements that prove that the numbers on your returns are correct will be necessary if you are audited. Those who own a business or have complicated tax returns will need to keep evidence of their financial actions and accounts for longer, and should check with a tax professional.

TIP: Many store receipts are now created with heat printers. These receipts fade and become unreadable over time, especially when exposed to heat. If you need to retain store receipts as part of your taxes, you will want to create digital copies so that they are still readable and available over time.

Other Bills and Receipts

This is an area where people need to use their own judgment and consider their own needs. Many people do not need to keep bills and statements for things like utility payments. Some people may keep them for a period of time to confirm that payment went through or to track utility usage—although this latter option is now sometimes available directly from utility accounts when you log in. If you run a business out of your home, you may need to retain utility bills as part of your proof of costs for business use of a home.

Other bills and receipts that you may need to retain include bills that you are in the process of contesting with your credit-card company or other firm; receipts for items that have a warranty; and receipts for large purchases. Keeping receipts for large purchases can assist with a legal or insurance claim. For example, homeowners and flood insurance providers are more likely to cover your losses if they know the age and original cost of your belongings. In cases of theft, keeping receipts may also help you to identify the brand and model of items that are no longer in your possession.

MEDICAL DOCUMENTS

For better or for worse, medical records are increasingly digital. Some doctors have embraced this more than others, and you may find that all, some, or none of your medical records are available online. This may mean that you can log in to an account and track your own health details, as documented by your healthcare provider, including following your recorded blood pressure and weight, viewing test results, and seeing past and current prescriptions. If you do not often change doctors and have this access, then you may not need much more in the way of saving and preserving your records. However, if you keep track of multiple medical histories (in the case of children, or caring for the elderly), then it is unlikely that you have one place to log in and track the details. In addition, it is always possible that those online systems will change or that you will use multiple doctors who do not all use the same system.

The reason medical records can be digitized is that they are records of tests and treatment: whether the item is in digital or physical form, those things cannot really change. You should be able to ask for a photocopy of almost any medical document related to your care because they are recorded facts rather than specific, one-of-a-kind documents. Even photocopies of insurance cards are usually accepted. Whether you choose to keep any physical records, there is a lot of value to keeping medical records digitally: they can be better organized, they can be searchable (in most cases), and, should you need them, they are easier to take with you when you visit a doctor. There are also a number of online tools and software programs available to help you with the process. However, as with all proprietary offers, keep in mind that you may lose access due to upgrades or obsolescence.

The Mayo Clinic recommends keeping the following information as part of a personal health record:[12]

- Information on allergies
- Medications, including dosages
- Chronic health issues
- Immunization records
- Dates of major illnesses or surgeries
- Living will or advance directives
- Test results

The following may also be helpful:

- Copies of current health insurance cards (in case of an emergency or lost card)
- Past prescription medications and dosages
- List of previous doctors, including specialists
- List of previous primary care doctors from whom additional records may be retrieved if necessary

If you change providers, do not assume that your medical history moves with you. Keeping your medical history yourself can assist you when you switch doctors without having to rely on your memory or the interconnectedness of medical institutions. If you have a limited medical history and few previous doctors, and much of your medical information is easily accessible through an online account, you may not need to keep or back up medical documents.

LEGACY PLANNING FOR YOUR DIGITAL DOCUMENTS

Digital legacy planning involves making a plan for your digital materials to allow people that you designate to control access to your digital items. As more people maintain their accounts and information digitally, it has become increasingly difficult for families to handle estates and get access to the accounts of their loved ones in case of emergency. More than accounts, often our photographs, genealogy, and other personal items are behind a password barrier. That barrier is there to protect us, but it can also exclude people who we would want to have access in an emergency. Rather than looking through an office for important papers (as was once the case), family members must now grapple with trying to hack computer passwords and guess what accounts a person may have had. In worst-case scenarios, accounts are not found or cannot be opened—leaving families without the access and possibly leading to the loss of memories and information forever.

Legacy planning for your digital media is similar to making a will: we may recognize that we should have a plan in place but don't because it is difficult and, in some cases, uncomfortable. However, legacy planning for your digital items is important, and can be a great help to your loved ones in case of emergency or death. Your business or organization may also lose access to documents that they need if you are unable to access your accounts. Alternately, you may want to ensure that some items are *not* available to others in the future—for example, you may not want personal letters that were kept for sentimental reasons to end up with family. You may also want your Gmail or social media account to be shut down in the event of your death.

Since many of us are reluctant to pass over a big list of our accounts and passwords to anyone (even family), we may avoid making plans. Thankfully there are a number of ways to ensure that your items are protected and that your wishes are respected. Your plan should include emergency preparations, but may also include sharing documents or photos regularly to key people to avoid loss. You will also need to consider that some things, like music and video downloads, may not be transferrable.

Below are a number of strategies for digital legacy planning. You may need to use multiple approaches, depending on the items you want to protect. Above all, no matter how you plan, make sure that you share your plan with family so that they don't spend unnecessary time trying to access items that do not exist.

Password Sharing

Any time you share a password, you are putting yourself at risk. However, some argue that providing loved ones access to personal and financial documents that you would want them to have outweighs the risk. As with will executors, roommates, and loan cosigners—choose carefully. Also, provide instructions: When is access permissible? Should access be available only after your death, or are there other situations that would warrant it as well?

You may choose to share your computer login password directly with someone you trust (perhaps someone who does not live in your city, as they would have limited access anyway). Do not email the password, and ask the person to store it securely and not identify it in writing or to others. Similarly, you can also give someone access by providing the computer login in a place where family or friends would look and have access to in case of emergency, such as a safe, safe-deposit box, or paper file. The challenge for all of these plans is the need to periodically change your password, which would require updating anywhere it is kept as well.

You could also keep a password-protected document on your computer desktop, or choose to password-protect other files. Adding passwords to your computer files can work in two ways: first, you can protect things like correspondence or personal items that you do not want available to your "digital executor." Doing so might allow you to give someone access to your computer login without fear that they will see some files. In this case, it helps to tell the person in question that any password protection is intentional, and that you prefer those files be deleted in the event of your death. Second, you can keep a file or document on your desktop with all of your account logins and passwords. This file can easily be updated because it lives right on your desktop, but it can only be accessed with the password. Consider giving the password to a key person in your digital legacy plan.

To password protect and encrypt any Microsoft Office document (Word, Excel, PowerPoint, etc.) on a PC, click on the File menu and select the Info tab. Simply enter a password in the Encrypt Document box and click Save. Previous versions of Word for PC offer an encrypt option by selecting the Office Button, then Prepare, and Encrypt Document (see figure 3.1). On an Apple computer, open the document and select Tools from the top bar menu options. Select Protect Document from the drop-down menu. Enter the password in the box and click OK. Type the password a second time and click OK again to confirm (see figure 3.2).

Figure 3.1. Microsoft office password-protection options on a Windows PC

File Sharing

Another plan to ensure access to files is to share them regularly. Beware: depending on how and where they are shared, files may not be secure. However, sharing multiple copies is the best way to protect personal items like family photos and video, which do not usually have many safety implications if they were to become public. You may choose to create shared folders on cloud storage like Skydrive and Dropbox. Multiple people can view these folders if you share them. You may choose to share widely and can do so even if others do not have accounts with these providers: links can be provided that allow access to anyone with a link. More ways to share documents and files, outside of the context of legacy planning, will be discussed in chapter 7.

You can also create shared or open albums on sites like Google Photo or Flickr. Photos can be set as public with downloading available, making them truly open and accessible. If that's too public and you prefer not to have photos on the open web, consider using Google Photos with private links. Google albums are technically open and available, but only to people who have the link provided by you. Family can even bookmark the link, making it possible to access items without a password or account.

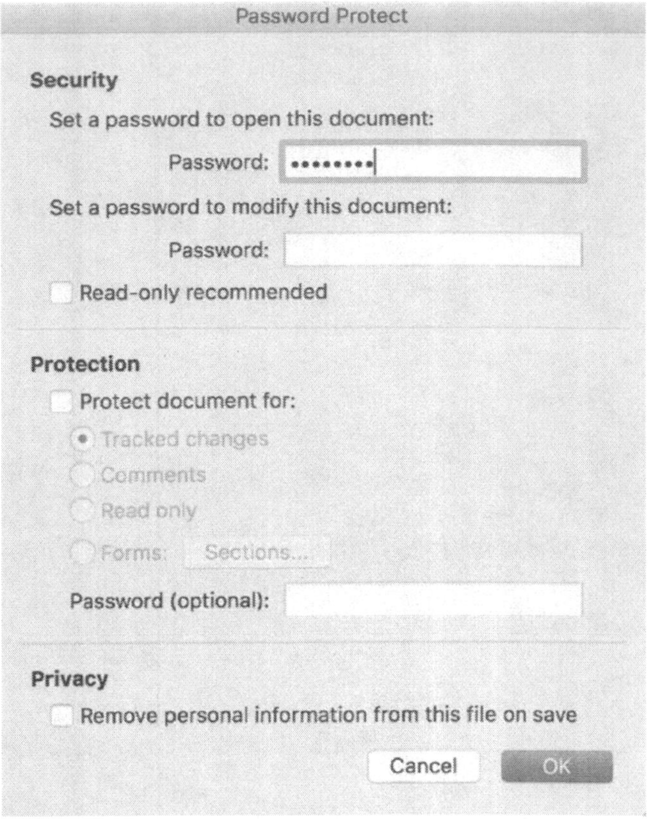

Figure 3.2.   Microsoft Office password-protection screen on an Apple computer

Planning for Your Online Accounts

- Email: Consider choosing an email provider based on what legacy options or restrictions are available. If you want to avoid family or estate executors attempting to access your email, choose an account like Yahoo that prevents any such access. To create special rules for your email after death, consider Gmail since it has the most options. If you check your current email system and don't like the policy on email access after death, switch accounts. More information describing what email options to look for is included in chapter 4.
- Social media: Make a list of the social media you use, and close any accounts you no longer want. Make legacy plans (see figure 3.3), including directions for whether accounts should be left up or taken down after your death. Keep in mind that some platforms will decide for you.
- Digital legacy planning: Look at services like those listed by Digital Beyond to find companies that email certain friends and family if you don't respond to

---

Legacy Planning Template [modify to meet your needs]

Accounts, Utilities, Medical & Legal
- I have provided a list of companies and accounts (not including financial) to [person's name] so that they can notify each company/cancel accounts in the event of my death.
- Access to financial accounts is included in my files and media plans.

Email
- I have used the gmail legacy feature to set up automatic deletion of my account after 6 months of inactivity.
- My Yahoo account has no legacy controls and will be deleted after a period of inactivity.
- I have provided [person's name] with a list of email accounts and my wishes that no one attempt to access them (allow them to be deleted).

Files & Media
- I have shared my computer login with [person's name], and will update them annually when I change my login information after the New Year. This individual has my permission to access my computer to get pertinent documents, including bank statements, family photos, and insurance information. Any files that are private and/or personal are password protected and access should not be attempted. I trust this individual to share and distribute my files as they see appropriate, and to delete any password-protected files according to my wishes.
- I have a document on my desktop called "If I die" which is password protected. I have also shared this document with [person's name] and the password does not change. The document contains:
  - bank account numbers
  - life insurance
  - retirement accounts
  - personal blog and domain name login information (& directions for each)
  - list of people I would like contacted if I pass away

Social Media
- I have designated [person's name] to be the executor of my Facebook memorial page using the Facebook legacy controls.

Figure 3.3. Sample legacy plan

their periodic check-ins. The idea here is that if you go "dark" for 60 days (or whatever you choose), your passwords can be emailed to a loved one.
- Account list only: If you are concerned about providing account numbers and logins (and that is a valid concern), consider giving key individuals a list of your accounts *without* numbers or logins. If you pass away, they should be able to gain access to pertinent accounts if they are legally allowed. Your list will give them the basic information so they don't have to guess or search for accounts.
- Provide family with a list of accounts that you prefer they not attempt to access.

REVIEWING WHAT'S IMPORTANT

After reviewing this chapter, you should be able to assess which of your important documents are at risk or in need of digital intervention. Using that information, you can now:

- Consider which of your legal documents require digital copies to protect yourself from loss during emergencies.
- Reduce the number of paper and digital files by discarding files that have passed their recommended retention dates.
- Reduce paper files by scanning some financial and medical documents that do not need to have physical copies.
- Encrypt and password-protect individual files to protect specific documents.
- Create a legacy plan by providing some form of access to your digital information in case of emergency.

NOTES

1. Federal Emergency Management Agency (FEMA), *Are You Ready? An In-depth Guide to Citizen Preparedness*, August 2004, https://www.fema.gov/pdf/areyouready/areyouready_full.pdf.

2. Leavitt Group, "Digitizing Your Important Personal Documents," *Leavitt Group: News & Publications*, June 3, 2015, https://news.leavitt.com/publications/digitizing-your-important-personal-documents/.

3. "Traveler's Checklist," U.S. Department of State, accessed April 1, 2016, https://travel.state.gov/content/passports/en/go/checklist.html.

4. FEMA, *Are You Ready?*

5. Mandy Walker, "How Long to Keep Tax Records and Other Documents," *Consumer Reports*, March 21, 2016, http://www.consumerreports.org/taxes/how-long-to-keep-tax-documents/.

6. Leavitt Group, "Digitizing Your Important Personal Documents."

7. FEMA, *Are You Ready?*

8. "Finding Unclaimed Life Insurance Policy Proceeds," Nolo Legal Topics Online: Wills, Trusts & Probate, accessed June 10, 2016, http://www.nolo.com/legal-encyclopedia/finding-unclaimed-life-insurance-policy-proceeds.html.

9. Walker, "How Long to Keep Tax Records and Other Documents," *Consumer Reports*.

10. "What Kind of Records Should I Keep?" Internal Revenue Service (IRS), accessed April 2, 2016, https://www.irs.gov/Businesses/Small-Businesses-&-Self-Employed/What-kind-of-records-should-I-keep.

11. "How Long Should I Keep Records?" IRS, accessed April 2, 2016, https://www.irs.gov/Businesses/Small-Businesses-&-Self-Employed/How-long-should-I-keep-records.

12. "Personal Health Record: A Tool for Managing Your Health," Mayo Clinic, accessed June 15, 2016, http://www.mayoclinic.org/healthy-lifestyle/consumer-health/in-depth/personal-health-record/art-20047273.

4

Correspondence

Email, SMS, and Voicemail

One of the most personal aspects of our digital life is the communication between ourselves and our loved ones. We also have lots of other correspondence, including work emails and texts, voicemails from friends and relatives, and information shared with larger groups. Personal communication has become increasingly digital because it is so much quicker and cheaper than it used to be. The number of email accounts has increased exponentially as letters mailed in the United States have declined.[1] The number of emails sent *each day* in 2015 was a staggering 205 billion on average, and that number is expected to climb.[2] Interestingly, although we have adopted digital tools for almost all aspects of communication, many people treasure the few pieces of paper communication that they receive.[3] That may be, in part, because the digital communication we use feels intangible. Many people do not save emails, texts, and voicemails because they don't know how—not because they don't want to. Thankfully, depending on what tools you use, most emails, text messages, and voicemails can be backed up to prevent loss.

Reasons for wanting to keep your personal correspondence are specific to each person. You may feel a great sense of loss if you lose texts when you switch devices or accidentally delete an important email from a loved one—even though you did not realize you were attached to those texts until they disappeared. On a more practical side, losing an email or text may hurt your business if it contains important information. When cell phone technology was in its infancy, many people experienced a less-than-optimal experience when switching phones, including the loss of saved contacts and messages. Email, too, is easier to save than it once was.

Options for backing up and saving will depend on the tools that you choose to use. Your options for backing up email are often specific to your email provider. Similarly, steps for backing up your short message service (SMS) text messages and voicemail messages may depend not only on your service provider, but your device.

For this reason, backing up is less straightforward than for other areas of digital preservation. Because services are different depending on the devices and the services you use, there is no solution that will work for everyone. This chapter will review two of the most popular email providers: Outlook and Gmail. It will also look at how to back up from Apple Mail's client application. If you are not using those providers, the chapter offers tips for where to find the information you need to save your emails. For SMS and voicemail, this chapter will offer suggestions for both Android and Apple devices. Additionally, this chapter will cover some basic ideas about managing your email inbox, voicemail, and texts more efficiently. If you find that you are currently using a phone or provider that makes backing up difficult, consider how important it is to you. If it is a priority, it may be time to switch to a new phone or provider.

TIP: If you don't care about your correspondence in the long run, then consider delet-ing unneeded messages regularly. This will save space on your phone and in your inbox.

Backing Up Email

This section will guide you through the steps for saving your email. Before get-ting started, consider that most email providers (especially the popular services like Gmail and Outlook) are very stable. You are unlikely to lose emails due to service failure from these services. Then why save them? Lots of reasons. First, you may not use these services forever. If you have important information that you know you will want to keep, back it up now so that you don't lose anything in transition. Second, you may run out of space. Many email providers have a limit on storage space. In an article in *Information Today*, Donald T. Hawkins reports that the two main mo-tivations for people to clean up their email inbox are full inboxes and leaving their job.[4] Ideally, your important emails will be safely archived and stored before you get a warning that you have hit your limit. Finally, though the major providers are relatively safe, they are also businesses. If you read the fine print of any free email provider, you may find that they are allowed to change the rules related to their service; this means they can impose space limits and discontinue tools even if you have used them for years. To be safe, protect any important items by backing them up elsewhere.

In general, there are three steps to saving or backing up your email:

1. Find out what backup options are available. Three of the largest providers are covered in this chapter (Outlook, Gmail, and Apple Mail).
2. Make inbox decisions, including how/whether organizing your emails into folders may help you when backing up.
3. Schedule backups. When will you do it, and how often?

TIP: Remember, you don't own your work email. If you want to back up your work correspondence in any way, remember to review your company's policies regarding email use and backup before proceeding.[5] *Also, when working with email at work, you may find that some of the settings described here are not available if your workplace has locked settings in your email setup.*

To Save or Not to Save

Before you start the process, do ask yourself whether it is worth the effort. If email isn't important to you, don't worry about saving it. Some people love to keep an empty inbox and regularly delete all of their emails (see the section on "Zero Inbox" later in this chapter). Before you do that, however, check the list below for some of the things you may want to stick in a folder for posterity and later use. These may be worth keeping, though you may choose not to save them anywhere but in your email account. If that fits your needs, then you may not want to do any additional backups.

Email messages you may want to keep:

- Shipping notices, until packages arrive.
- Receipts for online donations and other tax-related receipts.
- Confirmations of taxes filed online. These should be kept as proof that you filed in case you are audited.
- Receipts for large or expensive purchases. If you have an account with the company where you bought the item, the receipt is usually available there as well—but beware! Sometimes only recent receipts are available.

Step 1: Identify Your Options

The first step is to find out what options are available. However, to do that, you will want to identify whether you are using a mail client or webmail service. If you access your email on the Internet by going to a specific website (such as Yahoo.com), then you are using a webmail service. Webmail accounts can be accessed from any computer with Internet access. Mail clients are applications that are installed on a local computer for email management. Instead of using an Internet browser to go to a specific webpage, you open an application such as Mail (for Apple computer users) or Outlook (common on Windows PCs, especially in businesses). This is confusing for many email users, because many email providers offer both services. For example, Outlook users may use their client-based email when they are on their home or work computer, but may log in to Outlook webmail when they are traveling. Why does this matter? If you use a mail client on your computer, then your emails are already being saved locally. If your computer is set up to back up local files, then the files have already been downloaded from the Internet to your local computer, which has a backup. However, these can be hard to find since mail clients tend to bury emails in your computer. The location will depend on your operating system, mail client, and email setup, so it may take time to locate the files if you needed them. If you only use webmail, however, then you probably want to back up email. Similarly, if you are switching services, use multiple computers, or want to save emails in an open file format, then you may want to set up a backup system as well. Open file formats are not proprietary, so they can be opened in multiple applications. If it is available, the PDF format is the best option for exporting emails. MBOX is another common format for email exports. It can be opened with any mail application or with a third-party email reader application.

Backing Up Email from Microsoft Outlook

Outlook is one of the most common email services used in the workplace, and many people have it on their home computers. Outlook is available on PC and Mac devices as a mail client, and it offers an online version, Outlook Web Access (webmail). Backing up Outlook emails, either individually or in bulk, is much easier on the client version. Multiple methods for saving Outlook emails using the client version are included here. Outlook Web Access is not as full featured, and has limited save options. If you need to save from the webmail version, the best option is to print to PDF if that is an available printer option for that particular computer. However, if you are using the client version installed on your computer, backing up emails is remarkably simple. There are three main options: drag the email, convert to PDF, and the built-in archive.

First, Outlook has options for saving individual emails as PST (portable storage table) or MSG (email message) files. All you have to do is drag the email to your desktop or another location. This is easy, but both PST and MSG file formats are Outlook-specific. You may find some other applications that can open and read them, but in general using these files means that you will need a version of Outlook to open them.

If you are switching services or want to use an open format, there is a second easy method. Right click on any folder or individual email within Outlook and choose "Convert [file name] to Adobe PDF." As shown in figure 4.1, this method allows you to save one email or a group of messages in a format that is very readable. This option is great for saving a group of emails from a specific project or person, which is easily accomplished by gathering all of the intended emails into one folder.

The final method available to Outlook client users is the built-in archive function. This function backs up all emails in bulk so that they are saved elsewhere and do not take up email storage space. It is also automated, so you do not have to remember to do it. To set this up, choose File from the upper-right part of the screen in Outlook. Choose Cleanup Tools > Archive to see the options available (see figure 4.2). Options include choosing which folders should be archived, where the archived files will be stored, and how old items should be before archiving (see figure 4.3). You can exclude items from AutoArchive by checking a box in the individual emails (though this obviously requires some planning ahead), though you can override this function by checking a box in this menu. You can also set up AutoArchive controls from any individual folder by right-clicking on the folder and selecting Properties. Controls

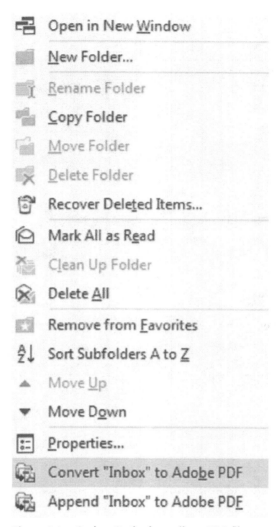

Figure 4.1. Saving Outlook emails as PDF files

are located on the AutoArchive tab. To find your archived files in Outlook, use the Go menu and look in the Folder List for Archive Folders. You can also find them from the File menu by clicking on Open and then on Outlook Data Files.

Backing Up Email from Apple Mail

Apple's Mail application is set up to pull email from another email provider. That means that the email provider may be Gmail, Yahoo, AOL, or another smaller service, but the email is downloaded to be stored and managed locally using the

Figure 4.2.   Outlook Archive, under Cleanup Tools

Figure 4.3.   Outlook Archive options

Mail program. Mail is only available on Apple/Mac computers. If you are using Mail, then you may have multiple options for archiving your emails: some may be built-in to your email provider's webmail access, and others are available from Mail. Looking at both options will help you identify which method is easiest and meets your needs.

Here are the steps for exporting an entire mailbox or folder from Mail:

1. Select one or more mailbox folders.
2. Choose Mailbox > Export Mailbox.
3. Choose a folder where the exported file will be stored and click Choose.

These directions are for Mail using Apple's Yosemite and El Capitan operating system, but earlier versions are similar. Files are saved in an MBOX file format, which can be opened in other mail programs, including Gmail and Outlook.

There are two easy options for saving individual files from Mail. Like Outlook, you can drag and drop individual messages to another location: simply select the file and drag it to your desktop or desired folder. This method saves the file in an MBOX format. To save files in a PDF format, select one or more messages and choose File > Export as PDF (see figure 4.4).

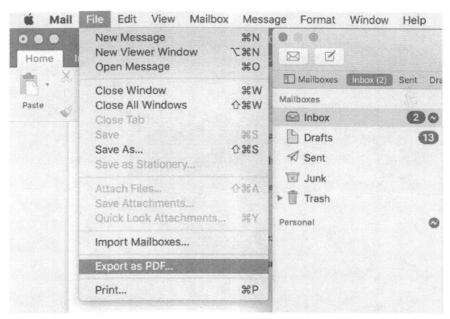

Figure 4.4.   Emails in the Apple Mail program can be easily saved as PDF files

Backing Up Email from Gmail

Google's Gmail email service is popular and offers options for backing up your correspondence. To get started, log in to your account online. To access the archive controls:

1. Go to the square of dots in the upper right and choose My Account.
2. Choose Manage your Google activity (under Personal info & privacy).
3. Select Control Your Content.
4. Click on Create Archive.
5. Select Mail (and anything else you want to back up) from the list (Figure 4.5).
6. You can "open" the Mail section to choose individual labels/folders (Figure 4.6).
7. Gmail will send you an email with a link. Click on the link to download the backup file.
8. The download file is a compressed ZIP file, which makes it easier to download.
9. Open the ZIP file to see the archived email file in MBOX format.

Figure 4.5.   Back up other Google files while backing up Gmail.

Figure 4.6. Choose which emails to archive based on labels.

Be aware that mail on its own is often very large, so it may make more sense to back up mail on its own unless you are just backing up individual folders or smaller parts of the inbox. You can easily save individual emails as PDFs using the print function from inside any email if your computer/network has a print to file or PDF option. Click on the tiny printer image in the upper right of the email. In the print-ready screen, go to File > Print and choose a PDF option from the print window.

Other Email Applications

Because of the large number of email applications available, it is impossible to list the save and archive capabilities that are available to users. To identify which options are available in your email account, open your email and look for an Options, Properties, Account Info, or similar heading. Go to your email provider's FAQ page (if available) or email the provider if you can find a Help or Contact Us email address. You may also be able to find an online tutorial for your specific email service by searching the Internet for "how to save email from [provider name]."

If your email service does not provide an Export or Archive option (or you cannot find it), you have a few options for saving individual emails:

- Use the Print > Save to PDF function, if that is available on your computer.
- Forward the email to another account with better options, where you can save or download the file.

- Copy and paste the email to another program (obviously not ideal for multiple emails).
- Take a screenshot of the email. This does not keep the text intact, since it turns what you see on the screen as an image. However, if you need to keep an email and have no other option, you can capture what you see on your screen. On a PC, hit the PrintScreen/PrtScn button. On a Mac, select Command + Shift + 3.

Step 2: Inbox Decisions

Now that you know your options, it is time to look at your email inbox. You may have noticed that some of the backup options inform how you might organize your email. For example, most backup options allow you to choose entire folders (or labels, like in Gmail). If you plan to back up some emails but not others, how you organize things may save you time. If you don't save many emails and don't want a complex system, consider adding just one tag/folder called Save Me. If you plan to archive emails and then delete them from your account, you might consider adding a date to each folder (such as Save Me 2016) so that you have a clear demarcation for what has been backed up and what has not.

Consider how you will use or search for the emails later. If you plan to export emails as a PDF then it may make sense to separate each PDF by what is included. Having a folder called Family Emails 2015 or ROI Project 2016 will help you export those folders in their entirety into a file that will be easier to use; if the emails are all mixed up, they will be harder to search for and read later. See chapter 2 for additional tips on folder and file organization, some of which are applicable to email folder organization. Most notably—give folders clear names and tags. Easy naming will allow you to back up more easily. Labels that have functions may simplify your life: creating folders called "Purchases" or "Travel" can keep those things from bulking up your main inbox, as well as help you find those items during tax or trip time.[6]

If you don't want to spend a lot of time on organization, make sure you are taking a good look at your email account's automatic sort and search functions. If you want to keep all emails from Matt Smith, many email services have options that allow you to route all emails from Matt Smith to a Matt Smith folder. When it comes time to save, his emails are already there. Similarly, you can search for his email address when you are ready to save, select all emails, and move them to a shared folder. Both ways achieve the same end result.

Outlook email rules can be set up by going to File > Rules and Alerts/Manage Rules & Alerts. Then, select the Manage Alerts tab in the pop-up window to add new rules. In Apple Mail, click on Mail > Preferences and click on Rules. Many tutorials are available online for both of these functions.

To set up email filtering rules in Gmail:

1. In the search box at the top of your email inbox, click on the down arrow.
2. Enter the criteria you want to find (an email address will get all things To: and From: a person; key words will find all emails on certain subjects, etc.).

3. At the bottom of the search window you can select "Create filter with this search."

4. Clarify what you want the search filter to do (figure 4.7).

5. Click Create Filter to set up the search filter.

TIP: You don't have to keep an organization system in place all of the time. If you want to archive your emails, you can set up folders, search for emails to put into each folder, export, and delete. If you don't want to use folder organization the rest of the time, that is OK!

To Organize or Not to Organize

Here's something you may not have considered: you don't have to organize your email inbox. That's right. Researchers from IBM found that using a robust search tool to find things in your email account was more effective for refinding items than creating folders and organization systems.[7] Similar studies find that adding tags, such as those used in Gmail, helped in searching, but also made it harder for some people to feel "in control" of their inbox.[8] As mentioned in the previous section, searching right before archiving may work just as well for you. If you have spent an unreasonable amount of time organizing emails into folders (as so many people have), only to find that it has not helped you refind things, then allow yourself to identify email

Figure 4.7. Setting up Outlook filter rules for automatic email sorting

as one of the things that you do not need to sort. A caveat or two about this choice, though: it does mean you have to be good at searching to find what you need. Search will work better if you regularly delete unimportant emails. Regularly deleting emails also prevents running out of email storage space.

Inbox Zero

Technology writers and bloggers sometimes refer to "Inbox Zero" as the ideal email management system. Mashable blogger Zoe Fox calls it the "Holy Grail of digital life-styles."[9] The idea of Inbox Zero is to keep your inbox empty most of the time. That does not mean everything gets deleted. Instead, users can set up filters and rules to organize their mail and delete things often (and in batch) when possible. Once you learn to archive and save important emails elsewhere, you may find that the ideal of inbox zero appeals to you. Some keys for making the system work include unsubscribing from bulk mail (rather than just deleting it), answering emails immediately whenever possible, forwarding emails that are not your responsibility, and saving emails that require more effort to a "needs response" folder, separate from your email. Hundreds of online articles and numerous books are devoted to the subject.

Step 3: Scheduling Backups

The easiest to think about but the hardest to actually do, scheduling and remembering to back up email messages is the most important step. If you just wanted to backup one or two emails, you may not need to schedule anything. Many people have an occasional email that they want to save, which they do as needed. If you plan to keep all or most of your email backed up, then you might consider backing up every few months or annually. If you use a calendar application, consider setting up a quarterly reminder. You might also plan to back up the previous year's correspondence every January. The key to scheduling email backups is twofold. First, you want to avoid backing up the same documents more than once, if possible. It isn't the end of the world if you back up the same files in multiple email sets, but duplicates can make things confusing—especially if you are saving your archives as MBOX files; when you load the files into a mail application, you will have duplicate emails. Consider organizing messages into folders each year, or sorting emails by date before backing up (if your email service allows it). If you are trying to save all of the emails from your sister, consider keeping them in a new folder or using a new tag each year (for example, Jane 2015, Jane 2016, etc.). This will allow you to save the whole year as a set, and it may make searching files easier in the future.

TEXT MESSAGES

Text messaging, also called SMS, is a short communication tool that many people use to chat via cell phone or other device. You may have both SMS (Short Message

Service) and MMS (Multimedia Messaging Service) messages. As the name implies, MMS messages contain photos, video, or audio. Saving text messages is not for everyone. Texting is an easy tool to use, and many people use it so frequently for day-to-day activities that saving all of their texts is not practical or helpful: they may never want to look back over so many messages. For example, how many times have you asked someone to pick up milk on their way home? This may not be at the top of the list of "things I must save and protect" when it comes to digital material. Other people may decide to save every text as a log of their life and conversations. Most people fall somewhere in the middle; there may be a few conversations they want to save, but the rest can all go. It is up to you and your personal needs.

TIP: When considering backup options, make sure the tool you choose includes MMS if you want to save those multimedia parts of the message along with the text.

One thing to consider when choosing to backup text messages is that most of the basic backup options are intended for recovery backup when a phone is lost or damaged. That means that they do not offer as many features for saving text messages to be viewed or saved in other formats: the services expect that you will load them onto a new phone for viewing. That creates limits on how useful saving your text messages might be, given your needs.

TIP: Even if you don't plan to keep texts forever, you can also use these backup tips to save messages that could be lost if you switch to a new device.

Saving Text Messages on Android Devices

You will probably need to download a third-party app to back up your SMS messages in bulk on an Android device. One of the best-reviewed SMS backup applications for Android is called SMS Backup+. A free tool, SMS Backup+ backs up messages to Gmail, can back up new and existing messages, and allows users to choose how often backups occur. Since Android phone users are required to have a Gmail account, backing up text messages to that location may work for many users. SMS and MMS files are saved with any attached media (like photos and live web links) and will appear under their own labeled section of Gmail called SMS. From there users can choose to use Gmail's backup options to save the messages they want, or they can simply keep them in Gmail. They can email the messages to others or download any media from MMS files. If you do not want to back up to Gmail, you should explore other Android apps in Google Play by searching for "SMS backup and restore" or "text message backup."

When choosing an app, consider the following things:

- Does it get good reviews overall? All apps have some bad reviews, but the overall feedback should be positive. There should also be enough reviews to know that the application has been well tested.
- Does the app back up and store the messages on a platform that you already use?
- Does it include already received messages, or just new messages after you install the app?

Saving Text Messages on Your Apple iPhone

Apple has two built-in ways to back up text messages on the iPhone. Since they are both available for free, you may want to explore the two Apple products to see if they meet your needs before searching for a third-party product. The two main SMS backup options include backing up to Apple's proprietary service, iCloud, or backing up a device to iTunes (also owned by Apple). Both options are built for backing up in case of loss; they do not offer many features for searching or browsing the messages once they are saved elsewhere.

Whether or not you use Apple iCloud for all of your backup needs, it may be the easiest method for backing up text messages from the iPhone. You can turn it on by going to Settings > iCloud > Backup and turning backup to ON (see figure 4.8). The backup function will save SMS and MMS messages, as well as some of the basic data from the applications that come standard on the iPhone. Once saved, the messages and other data will be available for backing up a new device through iCloud. Keep in mind that iCloud is free only if you use under 5GB of space. If you start to back up photos, videos, and other items in iCloud, make sure you have a plan for additional backup costs before you run out of space. If you don't use iCloud for your main backup, it may nevertheless be a great location for storing your text messages until you sort and delete them. This method is also automatic, so you do not have to remember to do it.

The second Apple-sanctioned method for backing up messages is to use a computer with iTunes installed to back up the iPhone and its data. This feature is intended as part of overall phone backup in case of loss or as preparation for switching to a new phone. For that reason, it is not as helpful if you want to view the messages you have backed up: it is meant to store the data for loading back to a new device. If you have an iPhone backed up and want to load existing messages onto a new iPhone, then iTunes backup will work for you.

If you want to save just a few messages here and there, you can do that without as much effort as setting up full backups. You can easily email individual or small batches of texts to yourself from your iPhone. Go to your text messages and open a conversation. Click and hold one of the messages until the "More" option appears. After choosing More, select all of the messages you would like to include. Then, choose the arrow at the bottom right to forward the messages (see figure 4.9). Instead of choosing a phone number, type an email address into the recipient field. The messages will be sent to the email address that is entered. While this method is simple and quick, it is not ideal for large groups of messages. The messages lose related metadata (embedded information) in transit, including the name of the sender and the date and time the message was sent. Groups of messages that are sent together are also bunched together with only a carriage return between each message. Despite these drawbacks, this may be your best option if you want to save one or two messages.

Another quick way to save a message is to take a screenshot of your phone screen. To do this on an iPhone, simply press the Power button and the Home button at the same time. This will save the screen as an image in your phone's Photos application.

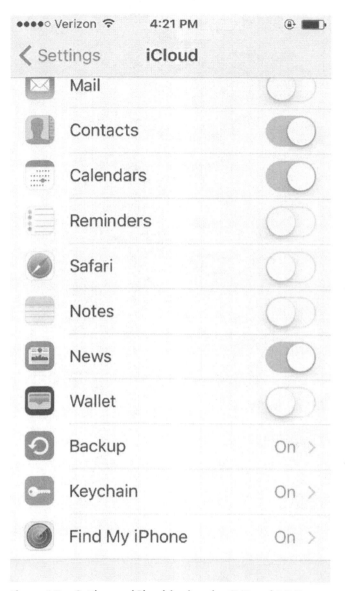

Figure 4.8. Setting up iCloud backup for SMS and MMS messages

Taking a screenshot will allow you to save the exact view of the screen, including the sender's name (at the top) and the date and time, if they appear on the screen. However, the image does not save any actual text since it has been turned into an image. That means it cannot be copied and pasted, or searched for individual words.

Figure 4.9. Select iPhone messages to include in an email

If you want to export a lot of messages and have more access to your text messages, or want to export and read them in another format (outside of an iPhone),
you will need to use a third-party app. Two of the most frequently recommended
applications for exporting text messages are also used for other file management:

AnyTrans and iMazing. These are popular applications, but they are not free. Many free apps exist, but offer a variety of options. Review the recommendation in the Android section for reviewing and selecting third-party applications before trying out possible tools.

Other Options

If your phone has limited options and you cannot use an application, there are a few basic features that may be used to capture text messages. All smart phones should have the ability to take a screenshot, as described in the last section for iPhones. Search for "Screenshot [your device name]" to find out how to take a picture of your screen on your specific device. Most phones, smart and basic, also allow you to email text messages by adding an email address into the TO: field rather than a phone number.

If you own a basic phone (a phone with fewer features, rather than a "smart" phone), you may be able to save text messages by removing the phone's SIM card (the identification card stored in the phone) and inserting it into a SIM card reader. SIM card readers cost $30–$35 and look like flash drives with slots for the SIM card. Since there are multiple types of SIM cards and multiple sizes, you will need to make sure to buy the reader that works with your specific card. Saving in this manner is usually more cumbersome than using an app. Some phones may take a mini-SD card, which also requires an adapter or reader that will allow it to be inserted into a computer SD slot. From your phone's text messaging section, you will need to find the option to save messages to card. Messages will be saved in a basic TXT file, and may lose formatting and other related data.[10]

TIP: Consider what you will do with messages after backup. Deleting them from your phone after transferring or backing up will free up space and avoid the extra work of backing up the same messages twice.

VOICEMAIL

Saving voicemail can be a tricky undertaking, depending on your device and carrier. Searching the Internet for ways to save voicemails results in many tips that are not practical, don't work, or result in poor audio quality. For example, playing your voicemail on speakerphone and attempting to capture the audio by recording on a different device usually results in very garbled, hard-to-understand audio. There are numerous free and cheap methods that work much better. Since there are so many options, this chapter will primarily focus on basic tools that will work for most users, regardless of device or carrier. The exception to that will be the Apple iPhone, which is in high use and offers several built-in options for saving voicemails. Other phones may have similar options available, and many third-party applications also claim to assist with saving voicemails.

Here are the basic steps for figuring out how to save your voicemail:

1. Identify whether your messages are saved on your device or a provider's server.
2. Identify whether your phone has built-in options or tools for saving voicemails or sending them to a location where they can be saved.
3. If you do not have built-in options, consider transferring/recording voicemails to a computer using an auxiliary cable and free software.

Step 1: Where Are They?

Before getting started with voicemail saving, identify whether your voicemails are saved on your phone or remotely by your provider. If you have to call in to get your voicemails, then they are not stored locally on your phone. If that is the case, you will want to check with your mobile provider to find out how long voicemails are saved. This may be listed on your provider's website, or you may have to contact the provider. Length of saving voicemails has no consistency, and you may find that voicemail is saved for only two weeks or one year. It is unlikely that your provider will save voicemails indefinitely: they prefer to have a policy in place that frees up space after a period in time. That does mean that you are working under a deadline! If you have an emergency, such as crucial information on a voicemail that is due to be deleted in a matter of days, you should contact your provider. Sometimes they can extend save periods or provide the audio files, usually for a fee. This is only the best option if other methods won't work for you and if there is a nearing delete deadline.

Step 2: Checking for Built-In Tools

If your messages are saved to your local device, that means that they are saved somewhere on your phone and can be extracted or sent elsewhere. In terms of preserving the audio files, it is much easier to move and store the files when they are stored on your phone. If you back up your phone, chances are that the files are saved . . . somewhere. That does not usually help you, however, since they are not always saved with easy-to-identify file names in a location that you can find without a lot of work. To see if you have any built-in options for voicemail saving, look at your saved voicemail for a menu, options, or archive. Many phones have options for emailing the message or copying a voicemail to a memory card.

Saving/Transferring from the iPhone

iPhone allows messages stored locally to be emailed or saved in Voice Memo. Many phones have similar offerings. To save your iPhone voicemail messages, which are stored locally on your device, go to Phone and choose Voicemail from the bottom menu. Select a message from the list, and it will open to show more options (see figure 4.10). Select the Export icon, which looks like an arrow rising out of a box. This will display multiple options for sending or saving the voicemail in another lo-

Figure 4.10.   **Export voicemails from iPhone using the built-in tool**

cation, including the option to email, save to Notes or Voice Memo, or send via text message. If you use Dropbox or similar online storage applications and the app is on your phone, you can also save the audio file for your voicemail directly to Dropbox. If you save to a new location (rather than mail or text), you can rename the file.

Step 3: Exploring Other Options

In the absence of a built-in tool on your device, you will need to explore options for transferring the audio files to a new location outside of your saved voicemail. One popular method of transferring/recording each message to a computer via auxiliary cable is described below. If this method is too complicated and you own a smart phone, then purchasing an application might be a better choice.

Audacity and Other Audio Editing Tools

Audacity is a free, open-source software tool used to create and edit audio files. It is certainly not the only tool that can help you save voicemails (and other audio). However, it is free, works with both Macs and PCs, and numerous tutorials for the tool are available online. To use Audacity to save voicemails, you will need to download it to your computer (it is not a phone app). You will also need an auxiliary/stereo cable, available at low cost in any electronics department.

To record voicemails:

1. Open Audacity and select Edit > Preferences > Recording (see figure 4.11).
2. Check the box next to Software Playthrough.
3. Plug your phone into your computer using the auxiliary cable.
4. Hit Record in Audacity, then play the message on your phone. If you must call in to your voicemail, hit Record in Audacity before calling.
5. When the message is finished, hit Stop in Audacity.
6. If you have extra audio before or after the voicemail, you can trim it by highlighting those sections and hitting the backspace key on your computer keyboard.

To find additional tutorials for Audacity, search YouTube or the web for "Audacity voicemail tutorial." Save files as WAV or MP3 files, since these are open formats that do not require specific tools or vendors to play. All computers (and most phones) should have a tool built in to play these formats.

Other Options

If your phone does not have built-in tools to send files elsewhere and you do not want to use audio software to transfer to a computer, consider one of the third-party applications specific to your device and budget.

Things to consider when purchasing a third-party application for voicemail:

• Does the app get good reviews? Have there been enough reviews to see that many people have used it successfully?
• What formats are available for saved voicemails? Look for MP3, M4A, or WAV, which are open formats that can be played on almost all computers and phones.

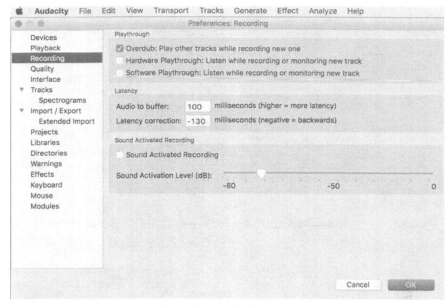

Figure 4.11. Use the free audio tools in Audacity to transfer audio

- Does it allow you to back up voicemails automatically? This is a big advantage over most built-in options, which are usually a manual process.

CORRESPONDENCE: AN OVERVIEW

This chapter has covered a lot of methods for capturing correspondence, though not all solutions will work for all people. Furthermore, technology changes quickly, requiring methods and tools to also adjust.

- For email: check with your email provider first to see what built-in tools exist to help you. When saving, choose open file formats like PDF or MBOX.
- For text messages: back up as part of backing up your phone, or explore third-party applications for more features.
- For voicemail: If your phone does not have built-in options for saving voicemail, you can connect your phone to a computer and use free audio editing tools to save voicemails. Alternatively, you can explore third-party apps using tips provided in the voicemail section of this chapter.

Overall, the thing to remember is that all digital correspondence is made up of computer files. Whether it is email, audio, or text, the related file is stored somewhere in order to provide access to users. By identifying where items are located and finding the right tools, digital correspondence can be saved in open formats in safe locations, ensuring that it stays available for years to come.

NOTES

1. "Pieces of Mail Handled, Number of Post Offices, Income, and Expenses since 1789," United States Postal Service, February 2016, http://about.usps.com/who-we-are/postal-history/pieces-of-mail-since-1789.pdf.

2. "Email Statistics Report, 2015–2019," Radicati Group, March 2015, http://www.radicati.com/wp/wp-content/uploads/2015/02/Email-Statistics-Report-2015-2019-Executive-Summary.pdf.

3. Rachel Rodriguez, "In E-Mail Age, Still Nothing Like a Handwritten Letter," *CNN iReport,* May 27, 2010, http://www.cnn.com/2010/LIVING/05/27/letters.irpt/.

4. Donald T. Hawkins, "Preserving Email," *Information Today* 31.6 (July/August 2014): 20.

5. Melanie Pinola, "How Can I Save All My Emails for a Personal Backup?" Lifehacker, March 14, 2013, http://lifehacker.com/5990556/how-can-i-save-all-my-work-emails-for-a-personal-backup.

6. Zoe Fox, "5 Tricks to Finally Achieve Inbox Zero," Mashable, October 10, 2013, http://mashable.com/2013/10/10/inbox-zero/#tiCdfnMz_Eqo.

7. Steve Whittaker, Tara Matthews, Julian Cerruti, Hernan Badenes, and John Tang, "Am I Wasting My Time Organizing Email? A Study of Email Refinding," *Proceedings of the SIGCHI Conference on Human Factors in Computing Systems,* May 2011, doi:10.1145/1978942.1979457.

8. Andrea Civan, William Jones, Predrag Klasnja, and Harry Bruce, "Better to Organize Personal Information by Folders or by Tags? The Devil Is in the Details," *Proceedings of the American Society for Information Science and Technology* 45.1 (2009): 1–13.

9. Fox, "5 Tricks to Finally Achieve Inbox Zero."

10. Mike Ashenfelder, "Archiving Cell Phone Text Messages," *Perspectives on Personal Digital Archiving,* March 2013, http://www.digitalpreservation.gov/documents/ebookpdf_march18.pdf.

5

Digital Photographs

According to the 2016 Internet Trends Report, approximately 9.5 billion photos are added to Facebook each month, and 2 billion photos are shared each day on the platform.[1] That is just Facebook. Because many people now take photos with their phones and tablets, photos are being created constantly. As Rose Eveleth notes in a 2015 article for the *Atlantic*, "Another way to think about it: Every two minutes, humans take more photos than ever existed in total 150 years ago."[2] The problem, of course, is that more does not necessarily mean better. Only a small number of photos that are created are shared, and even fewer may be organized or backed up for later use. Most end up dumped in files with their computer-generated names (see figure 5.1). Relatively few are printed. In fact, many photos will never be looked at again after they are first taken. In part, that is because it is so easy to take them, but decidedly *not* so easy to organize and manage them.

This chapter will outline what needs to happen to make all of that photo-taking worth it: how to evaluate, organize, and use your photographs in a meaningful way into the future. Here are the main parts of digital photo management:

- Managing the volume/evaluating what to keep
- Organization
- Handling metadata (data related to your photo file)
- Naming and labeling
- Storage
- Sharing

There are entire books about photo organization (some of which are listed in appendix B). This chapter cannot describe all options for managing photos. Instead, it

2015-06-25 16.55.12.jpg	Jun 25, 2015, 6:55 PM
2015-06-25 16.55.17.jpg	Jun 25, 2015, 6:55 PM
2015-06-25 16.55.38.jpg	Jun 25, 2015, 6:55 PM
2015-06-25 16.55.56.jpg	Jun 25, 2015, 6:55 PM
2015-06-25 18.29.23.jpg	Jun 25, 2015, 8:29 PM
2015-06-26 10.40.35.jpg	Jun 26, 2015, 12:40 PM
2015-06-26 10.52.58.jpg	Jun 26, 2015, 12:52 PM
2015-06-26 11.01.36.jpg	Jun 26, 2015, 1:01 PM
2015-06-26 11.01.50.jpg	Jun 26, 2015, 1:01 PM
2015-06-26 11.14.08.jpg	Jun 26, 2015, 1:14 PM
2015-06-26 11.15.34.jpg	Jun 26, 2015, 1:15 PM
2015-06-26 11.15.35.jpg	Jun 26, 2015, 1:15 PM
2015-06-26 11.15.38.jpg	Jun 26, 2015, 1:15 PM
2015-06-26 11.18.40.jpg	Jun 26, 2015, 1:18 PM
2015-06-26 11.18.41.jpg	Jun 26, 2015, 1:18 PM
2015-06-26 11.20.11.jpg	Jun 26, 2015, 1:20 PM
2015-06-26 11.20.20.jpg	Jun 26, 2015, 1:20 PM
2015-06-26 11.20.23.jpg	Jun 26, 2015, 1:20 PM
2015-06-26 11.20.35.jpg	Jun 26, 2015, 1:20 PM
2015-06-26 11.20.47.jpg	Jun 26, 2015, 1:20 PM
2015-06-26 11.23.38.jpg	Jun 26, 2015, 1:23 PM
2015-06-26 11.23.39.jpg	Jun 26, 2015, 1:23 PM
2015-06-26 11.23.47.jpg	Jun 26, 2015, 1:23 PM
2015-06-26 11.25.10.jpg	Jun 26, 2015, 1:25 PM
2015-06-26 11.25.28.jpg	Jun 26, 2015, 1:25 PM

Figure 5.1. A common photo folder, with last modified dates indicating that photos have not been changed (or likely looked at) since their creation.

will cover enough to get you started and should be of sufficient help to the average photo-taker.

TURNING DOWN THE VOLUME

We take too many photographs. That's good news, in some ways: you can take multiple shots so easily that one is sure to come out right. You can also have your camera phone with you at all times to catch anything that happens, and you do. Capturing things in our daily lives and during special events is important to many people. The issue is that, at the end of the day, there may be 1,000 photos on your phone, including duplicates and the blurry shots that missed their mark. When we used to get photos developed from film, it was common to go through them and throw away the blurry and accidental shots. It was also really unlikely that we took 10 photos of the same thing because we just did not want to waste the film. We probably also did not take photos of things we wanted to remember for later (for example, the brand of this fantastic peanut butter), as many people are now prone to do. Now, we have them all, and there is no set point (such as when the photos "arrive" in print) when we see them and make the choice to keep or discard. That is why the first step to dealing with digital photos is to institute some weeding.

Weeding your photos is one of the most important things you can do to improve their findability and relevance. Consider this: even if the only organization and labeling you commit to is putting photos into folders for the year they were taken, it is much easier to find a specific photo if you have weeded out those that are not really important. Having fewer photos—ideally all photos that you have consciously chosen to keep—makes it easier to browse them and find what you are looking for. It also makes them more special. When you have 10,000 photos from one calendar year, you and your family are unlikely to ever commit to sitting down to look at them. However, you may be interested in looking at the "best of" selection from that year. Your friends, too, are more likely to want to see the 20 very special photos you chose from your vacation rather than the 650 you originally took. Consider this especially when you are choosing to share photos: even if you personally want to keep more photos, a smaller number of specially chosen photographs is more likely to be viewed and appreciated by friends and family.

You must decide where the line is for what you will and will not keep, but in general it is necessary to get rid of photos you will not find important in the years to come. That means you need to be harsh. Judge your photos—then delete them (see figures 5.2 and 5.3). Here are general considerations when deleting your photos:

- Is it a duplicate? If you took eight takes of the same photo because mom kept blinking, choose the best one and delete the rest.
- Is it blurry? Unless it is the only photo of bigfoot or there is another compelling reason, don't keep blurry photos.
- Is the photo full of strangers? Frequently people end up with photos of other people's friends and family, or random people at an event. Or, in the case of old photos, perhaps you used to know those people but you cannot even remember their names. In this case, consider sending them on to people who might remember/appreciate them, and delete your copies.
- Is the subject too far away? There is a large difference between a selfie with Bob Dylan on stage way in the background and a photo of just the stage (usually in poor lighting) way in the background. Will you remember who was on stage when you look at that photo in five years?
- Is it too dark, or otherwise needs editing? It is true that you could use a filter to lighten a photo and it may turn out, but most people probably will not do that for most of the dark photos they have taken. If you have other photos from the same event or time, delete the dark photos or download an app and filter them right away.

Figure 5.2. This photo of a mansion was taken directly into the sun and is unlikely to turn out even after editing.

As with all matters of personal digital archiving, your choices will be personal, so not all of these tips will apply to you. Do your best, though, to ask yourself why you want to keep each photograph. This thought process is needed when working with a large number or backlog of unsorted photographs, and having some criteria before you start will help.

When and Where to Weed

There are a few basic approaches to how and when you can address the volume issue: that is, there are a few different points that make sense for weeding. If you follow suggestions laid out in earlier chapters, you should have all of your digital media being routed to one place for storage and backup. Either that, or you have all of your things backed up from the location they are in. If you take photos on multiple devices, the ideal option is to have them route to one central location: either an online location like Dropbox, iCloud, or Google Photos, or a local location like your home computer (check out chapter 1 for more information).

If your photo locations are all decided, the next decision is to choose where to weed them. You can either reduce them at the source—the place where the photos are created—or at the hub/end location. Both have advantages. By reducing at the

Figure 5.3. This blurry chair is one of many taken during a furniture shopping outing, but it was never deleted and now clogs up the photo folder.

source (such as digital cameras, phone, or tablet), you will delete any poor-quality photos before they end up getting stored forever, unnecessarily using up your storage space. The obvious way to do this is to go through your photos often on your phone or other device, but that is easier said than done. Consider these tips for weeding at the source:

- Vacation photos can usually be weeded on the way home while you are waiting for a flight in the airport, or on a plane. Take 15 minutes to go through and delete photos rather than watching a show on your flight or browsing the airport gift shops.
- Waiting in line at the grocery store (or anywhere) is a great time to delete photos from your phone.
- Any time you find yourself playing a game on your phone is a time you could spend weeding photos. This may not sound as appealing, but choosing to do it even once a month will help keep your photos under control.
- If you commute on public transit or you are not the driver, take one morning a week or month to go through your photos.
- Make a habit of deleting bad photos and extras right away.

You can also choose to weed your photos at the "hub" or end storage location. If your hub is a laptop or desktop computer, one advantage is seeing the photos on a larger screen. You can clearly see which photo is the best of a set, and you may have an easier time editing photos from the computer. Editing at the hub/storage location will also be necessary if you already have a huge backlog of photos that were never weeded.

Whether you delete unwanted photos at the source or at the hub, you will need to set aside some time to manage photographs if nothing else works. Depending on how many photos you take, you may want to set up some time weekly, monthly, or yearly. Consider setting aside a holiday or weekend morning to work on them. It is difficult to find time, but ask yourself: if I don't want to spend any time with my photos, why do I take them? If you find photos valuable, then they deserve a little bit of management time.

TIP: Think about printing a few photos once or twice a year. You will be able to identify some of the best photos from a particular period of time while you are going through them.

PHOTO ORGANIZATION

Once you have weeded out the photos you don't want, you may still have an overwhelming number left. Having an organization system will help you find photos later and store them in a way that makes sense to others who may need to look at your collection.

DO I NEED A PHOTO MANAGEMENT APPLICATION?

Photo management applications are very popular, and they offer a number of tools for editing, sharing, and organization. Using an application or program, either on your phone or computer, may help you manage your photos. A few of the most common choices are covered later in this chapter. However, apps can also complicate things. If you use a specific tool to manage your photos and that tool either changes or is discontinued, it can be difficult to work with your photos in the way you have previously. The best approach to using photo management tools is to have a basic organization system for your photos on your computer before using an application. Most photo applications import or "find" the photos you have on your computer, either by searching for image files or by using folders you specifically select. In other words, having a basic organization system will not prevent you from using management applications, but it will protect you if those applications stop working or change in a way you don't like. This does mean that phone applications may not be the best choice for organizing photos—though they are still great for editing them.

Creating an Organization System

The most important step in managing and organizing photographs is to get all of them in one place. This may be a "my photos" or "images" folder on your computer. If you back up to Dropbox or a similar online storage solution, you can set up folders there as well. Both online and local (physically on your computer) storage locations can work. However, there are a number of reasons to keep your photographs located on a local computer or drive rather than in the cloud:

- You can have access even if you are not connected to the Internet.
- Photos stored on your computer will be part of your backup system, which protects you from loss (see chapter 1).
- It is easier to take advantage of batch-renaming functions built in to computers (see chapter 2).

The most notable down side to storing photographs on your computer is that they take up a lot of space. Online photo storage can usually be expanded (for additional cost), but that is not as easy on a computer. If you want to store photos locally but need more space, purchasing an external hard drive may be the best solution. It will still be local, but can provide extra storage space close to home. External hard drives are much cheaper than buying a new computer, and may be cheaper than buying additional online storage space.

The next step in setting up an organization system is to set up folders. Date organization is the preferred organization system for photographs. Here's why:

- It is number-based. Using the year, month, day reverse date order (2016-08-01 or 20160801) will make your folders and files automatically sort in date order.

- It can be broad or specific: if you want things organized to the day or to the decade, date-range folders can adapt.
- Photo browsing makes more sense by date. Subfolders or individual photos can still be organized by specific events or named with keywords, making them easily searchable. This gives you multiple options for finding your photos.

Despite these advantages, you may have an organization system that makes more sense to you. If so, the important thing is to be consistent.

Setting Up a Date-Based Organization System

You can follow these steps to create a date-based photo organization system:

1. Decide where you will keep photos. The built-in Pictures or Photos folder is fine, or you can create a new folder in your Documents folder or elsewhere.
2. Start by creating folders, with the plan to move and sort later. It will help to have the structure set up before you start sorting.
3. Create broad, date-based folders inside the main folder. Start with decades (2000s, 1990s, 1980s, etc.). Make sure to name the folders with the date in front so that they organize in order.
4. If you have a lot of recent photos, create folders for individual years within the most current decade (and older decades as needed). Again, name folders with the year so that they organize in order: 2011, 2012, and so on (see figure 5.4).
5. Create a To Sort file in the main photos folder. This is exactly what it sounds like: a folder of photos that have not yet been sorted into year. This will allow you to put photos somewhere until they are sorted without cluttering things up. Sometimes that visual "clutter" makes organizing feel overwhelming.
6. Once this basic structure is in place, start moving your photos from all other computer locations to the Photos folder: either to the To Sort folder or directly into decades and years. When sorting photos into folders, it is extremely helpful to open two Windows Explorer (on PC) or Finder (on Mac) windows. This will allow you to easily drag and drop photos into the right location (see figure 5.5).
7. As decades or years fill up, and if date information is specific enough, create additional folders for months and dates, if you wish. Month folders should start with the numerical month (including the 0) in front so that they organize correctly. For example, 01 January or 201501January. (Note: unlike file names, it is usually just fine to have folder names with spaces in them).
8. Within each year or month you may wish to create folders for specific events or trips. If you have months and/or individual photos organized by date, consider labeling the event folder starting with the numerical month so that it sorts with the other folders. For example, 05May_CindyWedding or just 05 Wedding.

Figure 5.4.   Organize photo folders by year so that they organize chronologically.

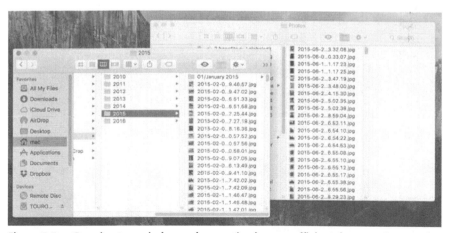

Figure 5.5.   Opening two windows when sorting is more efficient than copying or using the move function.

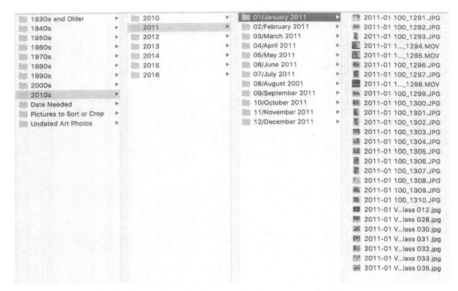

Figure 5.6.   Multiple layers of a photo organization system by date

Figure 5.6 shows multiple layers of a photo organization system by date. Remember, this is scalable. If dragging photos into folders for individual year folders works for you, then you may not need months or event folders. This is especially true if you name your photo files with dates: that means the whole year will be automatically in date order. Some cameras and phones label photos with the date, which can be a big help in keeping things organized.

TIP: This same date-based method can be used to sort physical photos. Create boxes and envelopes for decades, years, and months, and sort photos into each. This is a great way to prepare physical photos for later digitization. More on this topic can be found in chapter 8.

WHAT IS METADATA AND HOW CAN IT HELP ME?

Metadata is the data related to an item: both the technical details like file size and type, as well as the "about" details, like subject tags and file name. Some metadata is created and stored in your photo automatically at the moment that a photo is taken. This includes technical photo specifications, file type, file size, and sometimes (depending on your settings and device) geographic location. Metadata also includes file name, and possibly tags and ratings. Some metadata can be changed by the user, while other metadata (file type, location data) is traditionally read-only.[3] Archivists and librarians have written hundreds of books and articles about the use of metadata in photo archives and digital libraries, but much of that information is not useful to the everyday photographer. Most users just need to understand the basics: how metadata can help them, and which properties they can control.

File Types

One of the metadata properties that you may be able to control is the file type. As with other files, using open or common file formats is important. Open file formats can be opened on any computer with multiple different software applications, as opposed to proprietary formats that require a specific program. For photographs, open formats like JPEG and TIFF protect your photographs from software unavailability down the road. Most camera phones create photos automatically in JPEG, which is an open file type. TIFF (Tagged Image File Format) is another file type that many photo enthusiasts find helpful. TIFF files are not compressed like JPEG, which means they usually have better image quality and can show more detail.[4] However, that means that TIFF files are also larger. If you are interested in taking higher-quality photographs using TIFF and are willing to make extra space available to store the files, you will need to download a phone app or use a digital camera that gives you this option. On phone cameras you will also have limited options for output quality. If you do have the option to adjust for higher-quality images before you take a photo, remember that a higher number of pixels per inch (ppi) will allow you to print larger images from the digital file. About 300 ppi should be sufficient for printing photographs and most enlargements.[5] You can see how many pixels per inch an existing digital photo has by right clicking on the file and selecting Properties (on a PC) or Get Info (on a Mac): the pixels will be listed as dimensions or ppi.

TIP: You may have programs that can change, export, or "Save As" other file formats. You can use these to change or duplicate photos from TIFF to JPEG if you want a smaller file. However, saving from JPEG to TIFF cannot create a better-quality image. The quality was never captured in the JPEG, so changing the file type will not help.

Location Data

Many digital photos contain location data. Before the global positioning system (GPS) became popular, digital photographers had to add location metadata (if desired) manually. Now, however, camera phones and many digital cameras use embedded GPS to insert the correct location of the photo into the photo file.[6] This can be both helpful and problematic: having exact location is desirable for many users, but location can create privacy and safety issues. For example, consider whether you want the exact location of your vacation home or the playground where you take your children to be available to everyone when you share a photo online—especially when digital photos can be so easily shared.

TIP: Facebook and many other sites may strip some or all metadata from photographs to preserve privacy.[7] *The rules for each sharing website vary and change often, so pay attention to privacy settings or remove location information from photos if this is a concern.*

Digital cameras with GPS options will need to have the GPS turned on in order to capture a photo location. On camera phones, location data will need to be turned on. For iPhone users, location services can be turned off by going to Settings > Privacy > Location Services > Camera option Off/On. On many Android devices, the setting can be found in the Camera app by selecting the Location icon and setting it to Off/On. For other phones, you should be able to easily find your location services settings through a basic Internet search of "location services camera settings [phone make and model]." Keep in mind that once the location settings are turned off, location will not be captured and must be added manually if desired later. This usually requires a photo management application. However, if you want to remove existing location information from your photos, you can do that without any special software.

To remove existing location information from a digital photograph on a PC:

1. Right-click on the photo and choose Properties.
2. On the Details tab, scroll down to see the GPS coordinates (see figure 5.7).
3. Click on "Remove Properties and Personal Information" at the bottom of the window.
4. You will have the option to remove the properties from the file or create a duplicate with removed properties.
5. To remove details from the file, select "Remove the Following Properties from this File" and scroll down to GPS. Select Latitude, Longitude, and Altitude, then click OK.

To remove existing location information from a digital photograph on a Mac:

1. Open the photo in Preview.
2. Go to Tools in the menu and choose Show Inspector.
3. Select the "I" to get to the information panel and select the GPS tab.
4. Click on the Remove Location Info button.

Tagging and Keywords

There are several ways to add keywords or tags to photographs. There are built-in file management tools available to both PC and Apple users, and additional applications can assist with keyword and tagging management. Before getting started with tagging, consider how you will use your keywords. What are you likely to search for? Frequent tags include names, places, and events. Consider creating a naming system before you get started. For example, if you want to tag photos with the names of people who appear in each one, you should decide how you will list those names. If you choose to use "Mike Thompson" as a tag, you must remember not to use

Figure 5.7. Photo GPS location data shown on the Windows Properties Details tab

"Michael Thompson" next time, as the computer cannot tell that is the same name/person. Also, you may want to consider using separate tags for first and last names: so you might use "Nevaeh" and "Jackson" separately for Nevaeh Jackson. This will help if you want to search for photos including any or all members of a family who have the last name in common. It may also assist you when you cannot remember the spelling of a complicated name. For places, consider whether you will use city names, state names/abbreviations, or both if you plan to tag based on location. In general, the most important thing about tagging is to be consistent. If it helps, write down tags in a notebook or create an index document. Also, remember that keywords can be used in tandem with file names to help you when finding a specific photo or set of photos; you may not need to repeat the same words in both name and keyword.

TIP: You may be able to search by location without tagging if your photos have location data from your phone or camera. If they do, programs like Google Photos and Photoshop Elements let you search by location without any added work/tagging. See the section on photo management software later in this chapter for more information.

There are tools for adding keywords to photos on most computers. The basic ways are described here. Both methods will create keywords that will transfer between computers, including different operating systems (PC vs. Apple/Mac).

On a PC, you can add keywords to an individual photo using these steps (see figure 5.8):

1. Right-click on the Photos file and select Properties.
2. Click on the details tab to see description options.

Figure 5.8.   Adding keywords to a photograph on a Windows PC

3. Click next to Tags to add keywords.
4. Click Apply and OK to save.

These are the steps for adding keywords on an Apple computer (see figures 5.9 and 5.10):

1. Open the photo.
2. Click Command (Apple key) + I, or go to Tools > Show Inspector.
3. In the Inspector window you can see four icons across the top. Choose the third icon (the magnifying glass).
4. You can see any existing keywords or press + at the bottom to add keyword tags.
5. Tags are saved immediately. Close the window when finished.

Once tags are added, you can use them immediately. Using a file search on a PC or Spotlight on a Mac, simply search for a tag you have used and it should appear

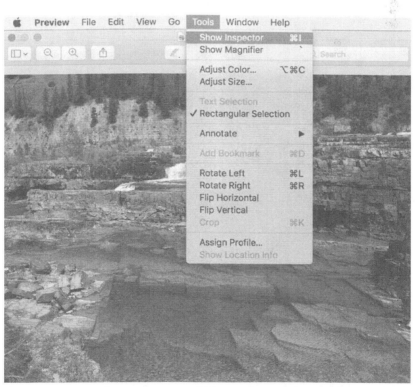

Figure 5.9. Open the Inspector to add keywords to a photograph on an Apple computer.

Figure 5.10.   Add keywords to the Apple Image Inspector screen.

in the search results (see figure 5.11). If you have moved all of your photos to one folder or location, searching just that location will improve your results, because text-based files using the same words will not appear. Tagging can be extraordinarily helpful when managing thousands of photos because you can search and find things quickly. If you are interested in tagging a large number of photos at once, consider using photo management software, described later in this chapter.

TIP: Mac users may easily be confused by the "Add Tags . . ." field shown in the Get Info window. The field appears when you right-click on any file, including a photo, and choose Get Info (see figure 5.12). This tags field is a feature of the Apple operating system that is intended for color-coding or tagging files for use on your computer: items like WORK and PROJECT123. These tags are not embedded into file data like photo keywords described above.

Desktop
NinjaPanda
Add a search filter
Date modified: Size:
Search Everywhere
2016-07-09 11.22.56.jpg
Tags: NinjaPanda
Type: JPEG image
Date taken: 7/9/2016 11:22 AM
Size: 1.98 MB
Search again in:
Libraries
Computer
Custom...
Internet

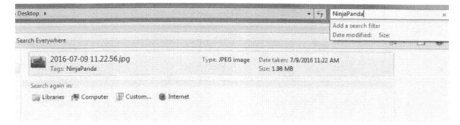

Figure 5.11. Searching for the keyword "ninjapanda" will bring up any item with that keyword added or in its title.

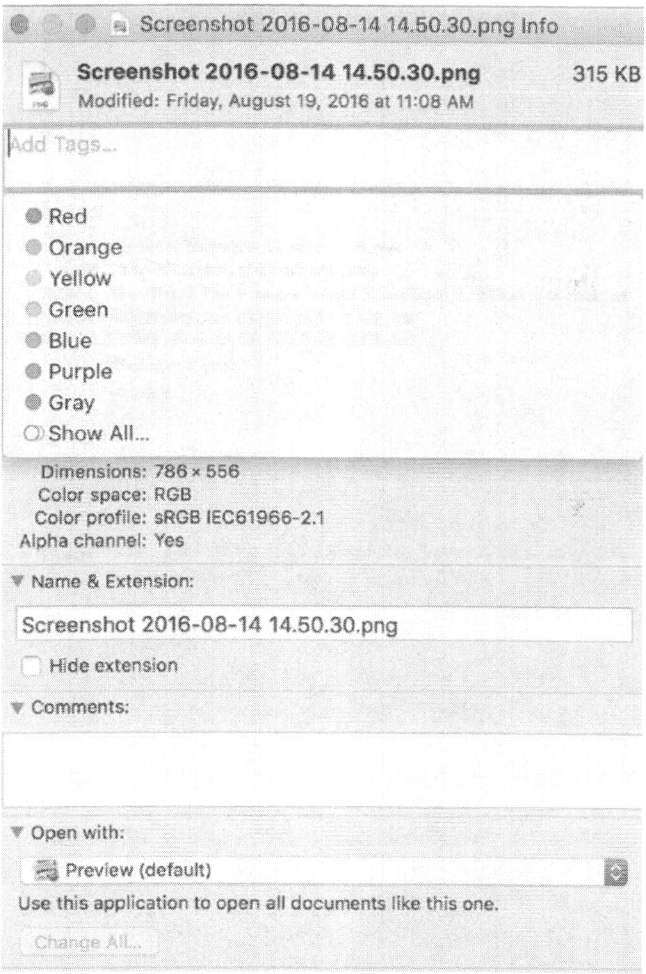

Screenshot 2016-08-14 14.50.30.png Info
Screenshot 2016-08-14 14.50.30.png
315 KB
Modified: Friday, August 19, 2016 at 11:08 AM
Add Tags...
Red
Orange
Yellow
Green
Blue
Purple
Gray
Show All...
Dimensions: 786 × 556
Color space: RGB
Color profile: sRGB IEC61966-2.1
Alpha channel: Yes
Name & Extension:
Screenshot 2016-08-14 14.50.30.png
Hide extension
Comments:
Open with:
Preview (default)
Use this application to open all documents like this one.
Change All...

Figure 5.12. File tags mean something different on Apple machines.

Metadata Standards: What You Need to Know

When working with tagging and other forms of photo metadata, you may come across references to EXIF, IPTC, and XMP. EXIF (Exchangeable Image File format), IPTC (International Press Telecommunications Council), and XMP (Extensible Metadata Platform) are metadata standards—three ways that photo information is embedded within the photo file.[8] For the most part, you do not need to know much about the standards to make them work for you. However, there are two things you should keep in mind: first, if you are exporting photos from one system to another and you have the choice to include or export EXIF, IPTC, or XMP data, you will want to say "yes" if you want tags and other information to transfer with the files. Second, because different photo management systems use different standards, keywords and photo data may not transfer between all systems, or they may transfer but be called something different (tags vs. keywords, or caption vs. description). Because metadata interoperability is improving, this may not be much of a problem, but it is something to look for if you are switching systems. If possible, test a new program to see if your existing tags transfer before switching wholesale to a new system.

Naming Photographs

How you name the individual files can help you while browsing and searching. This can go hand-in-hand with the file organization system described earlier, and can be used in tandem with keywords. Photos are unlike other files, and may require separate naming conventions. For example, most computers can search the text written inside text-based documents, making keyword searching more efficient. Photos do not have that option, and more descriptive names or tags may be needed. Nevertheless, file-naming conventions outlined in chapter 2 still apply: try not to have excessively long file names, and avoid special characters. That means that listing each person in a photo in the file name won't work; names may work better in a tag or keyword field.

What is the best use of photo file names, then? That is really up to you. Like tagging and keyword, location and event are common things to see in a photo name. Since you can add unlimited tags for more detail, using the file name for more broad subjects that apply to a group of photos may work best. Consider using date (specifically YYYY-MM-DD) at the beginning of the file name so that you can browse photos easily by date. While it is true that many photo organization programs organize photos by date based on the associated file creation date, those dates can be incorrect. For example, if you digitize older photos, they may have the digitization date rather than the date the photo was taken. Similarly, your phone or camera may use the date that photos were transferred to your computer rather than the photo date. If you have mainly new photos taken with a smart phone or date-capable digital camera, then you may not find a date prefix as helpful.

Once you decide on what to use in your photo file names, you can use the batch file-naming options described in chapter 2 to add text to a set of photo names. Since photos (like other files) cannot have duplicate names in the same folder, you may decide to keep a number at the end of each photo if you want them to share the description information. For example, you may choose to add 201607_BoulderCO_ to the front of all photo names from your recent trip to Colorado. If your camera assigned a random number as a file name, you can easily add the prefix to the front to better describe a set.

PHOTO-MANAGEMENT APPS

When considering the management of your digital photographs, you may decide to use the basic file-management tools built in to your camera and computer. If you do not often edit your photos and do not take (or keep) a large number, then you may not be interested in a photo-management application. If so, you will still be able to take full advantage of the basic functionalities described earlier in this chapter: Spotlight on Apple for updating file metadata; and Windows Photo Viewer Properties. You can also easily batch-change file names for groups of photos. If you find that these basic tools are not enough—for example, you want the opportunity to batch-change keywords, create shareable albums, or use facial recognition—then you will want to explore photo-management applications.

Photo-management applications have a number of advantages and disadvantages. The advantages include the many tools that are offered, including color editing, batch editing, facial recognition, easy online-album creation and sharing, and more. Each user will need to seek out the software that offers the tools that he or she wants, while also giving some consideration to ease of use and cost. Another concern is whether photos will be moved, changed, or duplicated from their original location. If you already keep your photos organized in files and want to protect yourself from future disruption if a software application becomes unavailable, look for a system that does not alter the main copy of your photos. Conversely, if you want to use a management program to help you organize, rename, and tag your photos, then you will want to make sure that any changes you make will "take" and remain with your photos even once you stop using the application. Answers to these questions can usually be found in a software website FAQ section. These concerns are addressed later in this section for some of the most common photo-management applications. Keep in mind, the applications described here are just a few of the hundreds available on the market. Decide what kind of features are important to you, and choose a program based on your personal priorities.

Photo-management applications fall into two main groups: desktop applications and online applications. Some online applications may require you to download some software, but in general a desktop application is one that you can use on your computer even when you are not connected to the Internet. There are also numerous apps for phones and tablets. However, unless you keep all of your photos on your phone or tablet (which is not recommended in order to avoid space issues and for ease of backup), it makes more sense to try to collect all of your photos in one location for sorting and management. That means choosing either a central hub location, like a home computer, or an online application like iPhoto. If you choose an online application, look for ways to back up your collection: you don't want all of your photo eggs in one online basket, no matter how safe it claims to be. Some online offerings allow users to upload a collection from their computer. If the core storage location is on your home computer but you organize and manage photos online, that may be a solution that addresses both management needs and reliable storage.

Photos for Mac Users

Apple's Photos (formerly iPhoto) is an application that comes standard on Apple computers, making it a free choice for some people. If you are a Mac user and you use iCloud for photo backup for all of your devices, Photos will be able to access and import all of your photos for use in the program. Photos organizes by date and has a viewing screen similar to the one that iPad and iPhone users see. Photos has options for organizing, sharing, and creating albums, as well as marking your favorite photos and facial recognition. It also has basic image-editing features like cropping and color adjustment, available by selecting a photo and choosing Edit in the upper-right corner. You can also edit the photo metadata (including keyword and description) from within Photos by right-clicking a photo and accessing the Get Info option. Photos stay in their original location, and updates to photos are made on the originals. Based on its price and interoperability with other Apple products, Photos may be a good management application for some Apple users.

Adobe Products: Photoshop Elements, Lightroom, and Photoshop CC

Adobe is a well-known name in digital design, and offers a number of applications for photo editing and management. It offers products for both Mac and PC devices at a variety of prices. The most well-known product, Photoshop, is now a subscription product, Photoshop Creative Cloud (CC). For about $10 a month, Photoshop CC offers a lot of tools for editing and creating, though not many tools for managing and organizing. Photographers interested in manipulating or fixing issues in their photos may be interested in Photoshop CC, which comes with a full version of Adobe Lightroom. Lightroom is a photo-organization tool that offers batch-import tools, slideshow creation, web-sharing options, extensive color management/correction, and sorting options. Unlike Photoshop CC, Lightroom can be purchased as a $140 download if you do not want to purchase the monthly subscription package for both products.[9] Lightroom also allows users to adjust and add photo keywords in XMP, which can be recognized by many other applications. Photoshop Elements is a less full-featured program than Photoshop, for a lower price tag. The latest version of Elements (as of 2016) costs $99 for a download version. It is geared toward casual (rather than professional) photographers, so it is more user-friendly and includes the most popular editing and sharing features, including facial recognition. With Elements and Lightroom, users choose specific folders of photos they want to include, and photos stay in their original location. When editing and adding metadata to a photo, you have the choice to save the updated photo as a new file, or you can overwrite the original with the new photo.

Google Photos

Google Photos replaces Picasa, a popular photo application that was discontinued in 2016.[10] Unlike Picasa, which was a program download to your local computer, Google Photos requires users to upload all photos to the web. This may be a deal-breaker for many users, though Google claims that there are robust privacy settings in place.[11] If you do choose to upload your photos to Google, you can do so from your computer, phone, and tablet using free applications. Once the photos are available in a Google Photos online account, the search and sort capabilities are very impressive because they take advantage of Google's search tools. Google's full-search algorithm can assist with photo sorting. Searching for tags, metadata, or words in a photo title will bring up the photos that apply. More impressive, searching for things like "cat" or "R2-D2" will find photos that apply without any tags or descriptions: Google can recognize those things in your photos without help (see figure 5.13). Keep in mind that the algorithm is not perfect, and may miss some results based on subject searches. However, searching for locations will almost always find photos with geo-location information in the photo—that's all photos taken by a mobile device, unless you have turned off location services. By entering Google Photos and clicking in the search box, you will also see a screen of available faces. Labeling the faces will allow Google to find more photos of the people you name by applying facial recognition.

While uploading photos to Google Photos is easy, it is an added step for some users. It gives you control over which photos are uploaded, which means you can choose to only upload photos you want to share. If you have thousands of photos to organize and sort, however, it may not be the right tool; changes to the photos in Google Photo do not change the originals on your computer, and there is a limit on how much space Google will give you for free. Since it is free and the search capabilities are strong, you can try it out, and may find that adding storage space is worth the cost.

Photo Storage

How and where you store your photos is also important. You may already have this step covered if you keep your photos primarily on your computer and have a backup system in place (see chapter 1). If you cannot or do not want to store all of

Figure 5.13.   Google Photos can recognize things in your images, even without key-words.

your photos in this way, there are other options. Unfortunately, not all of them are great options: you cannot save photos to just any media and stick these in a drawer and expect them all to be fine when you need them again—at least not over the long term. The Library of Congress refers to this as "store and ignore," pointing out that all hardware and software become obsolete and all storage media has a limited lifespan.[12] The first step to fighting obsolescence has already been covered: using open file types will reduce your reliance on specific software, which may become obsolete.

If you are already using open file types, hardware and storage are the main culprits in making your files inaccessible in the future. All storage media fails eventually, from CD to computer hard drive. However, photo storage media are not created equal. Some storage media have longer lives and lower fail rates—that is, your data is less likely to become corrupt or unusable, or the media is less likely to stop working.

Here are the keys to successful photo storage:

- Keep multiple copies (ideally three or more). This can be done as part of your overall computer backup system, explained in chapter 1.
- Do not depend heavily on stagnant storage—that is, don't rely on storage options that cannot be updated and are not in frequent use. External hard drives, for example, have higher failure rates when they are not "exercised" regularly—that is, when you leave them in a box in the closet. Checking files periodically is helpful for recognizing data corruption or loss as early as possible.
- Do not use the same kind of storage for all of your copies. If you burn three copies of your photos to CD, that may feel safe. However, all three are vulnerable to the same problems and may all fail based on one issue, like UV exposure or heat.
- Replace your storage media at regular intervals, or consider using online storage as part of your backup system. External drives should be replaced every three to five years, and flash drives can last five to ten years under ideal conditions.[13]

CDs/DVDs, Flash Drives, and Memory Cards

A CD (compact disc) and a DVD (digital video disc/digital versatile disc) are not good options for long-term storage, but they are useful for sharing and transfer of photos. Both types of optical media are susceptible to failure, especially when they are not kept in ideal conditions. For years after the increase in digital photo use, many photographers received or stored their photos on CD. Computers with CD/DVD burn drives made it easy to store photos this way. However, optical disc media—especially those burned at home—are prone to failure because of numerous vulnerabilities. Beyond the constant risk of scratching the surface, the coating on CDs and DVDs can flake off over time, especially when exposed to heat and extreme environmental conditions.[14] The quality of the disc also counts, as does the coating material and age. UV light and overuse of rewritable discs also contribute to failure.[15] Beyond data loss, some new computers are being released without CD drives, risking obsolescence of the media. Finally, in the case of non-rewritable discs, there is a practical issue: the data is stagnant, and cannot be changed or added to over time. That means that you cannot update or improve photos stored on those discs. Despite these disadvantages, optical disc media are easy to use (if you have a drive) and great for sharing. They do not require Internet access, so may be a preferred sharing method if online sharing is not an option. If you do choose to use disc media, make sure you have multiple copies of your photos, including some that are not on CD or DVD. Also, if you have old CDs that have not been copied to a computer and have no backups, moving those photos to a more secure storage medium should be a priority.

Flash drives (also called jump or thumb drives) and SD (Secure Digital)/memory cards are rewritable forms of portable storage that are also in high use among digital photographers. Most digital cameras and many computers have an SD or micro-SD slot so that photos can be transferred directly. Flash drives do not plug into most cameras, but are frequently used by many computer users for storage of photos and anything else they need. Both SD cards and flash drives are cheap and small, making them easy to get and store. However, they also have limited lifespans and are not as reliable as hard drive and online backup; both media last from five to ten years, depending on use, environment, and quality.[16] In general, only use flash drives, SD cards, and optical media when sharing rather than for archival or backup purposes.

Sharing

The best way to ensure a long life for your photographs is to share them. As with physical copies, sharing digital copies of your photos makes it more likely that someone will retain a copy if other storage fails. As discussed earlier in the chapter, many photo management applications have easy options for sharing albums and photos. However, not all shared albums give rights to your viewers. If you want friends and family to be able to download and save photos from your collections, make sure to choose an application that allows it. Google Photos has many options for sharing that allow the photo owner to set access privileges. Though they are not truly photo management sites, social media websites like Facebook may also feel like a place where family and friends can share your photos. However, these, too, are not usually accessible for download and may be reduced in size and quality for viewing on the site. Overall, the best way to share photos is to give other people a copy of your photo files. Storage options mentioned in the previous section—CDs, flash drives, and SD cards—are good options for sharing because they are cheap, require no special software, and are portable. Online-file-sharing websites like Dropbox allow multiple people to share files, including photos. Chapter 7 will cover this method of sharing in more detail.

TIP: When sharing photos, be aware that your photos may include metadata, particularly geo-location info: if that information is not stripped out, then anyone who has access to the photo will be able to find the locations where photos were taken. That may be fine for the Grand Canyon, but putting up photos taken at home makes it possible for people to know exactly where you live. Removing location metadata is explained earlier in this chapter.

DIGITAL PHOTOS: CHAPTER HIGHLIGHTS

This chapter reviewed some of the basics of digital photo management. Here are the highlights for review:

- Weeding photos regularly increases their value and findability.
- Using a basic date-based organization system can help you, whether you use a photo management program or not.
- Understanding basic photo metadata and using file-naming techniques can help you better manage your digital photos.
- Choosing the best photo storage media, making multiple copies, and sharing often will protect your photos into the future.
- Multiple photo management software applications exist to assist people with managing their photos, and many of them are free. Explore free and built-in tools and decide what capabilities you need in a program before purchasing.
- Sharing is the best way to ensure that your photos will be available to family and friends into the future.

NOTES

1. Mary Meeker, *KPCB Internet Trends Report 2016*, June 10, 2016, http://www.kpcb.com/internet-trends.

2. Rose Eveleth, "How Many Photographs of You Are Out There in the World?" *The Atlantic*, November 2, 2015. http://www.theatlantic.com/technology/archive/2015/11/how-many-photographs-of-you-are-out-there-in-the-world/413389/.

3. Metadata Working Group, "Guidelines for Handling Image Metadata, Version 2.0," November 2010, http://www.metadataworkinggroup.org/pdf/mwg_guidance.pdf.

4. Metadata Working Group, "Guidelines for Handling Image Metadata."

5. Butch Lazorchak, "Four Easy Tips for Preserving Your Digital Photographs," *Perspectives on Personal Digital Archiving*, March 2013, http://www.digitalpreservation.gov/documents/ebookpdf_march18.pdf.

6. Metadata Working Group, "Guidelines for Handling Image Metadata."

7. Sin Mei C, "Why Facebook and Twitter are Stripping Out Your Context," *Sentiance: Opinion*, October 11, 2013, https://www.sentiance.com/2013/10/11/facebook-twitter-stripping-context/.

8. Metadata Working Group, "Guidelines for Handling Image Metadata."

9. "What Is Creative Cloud?" Adobe Creative Cloud, accessed July 16, 2016, http://www.adobe.com/creativecloud.

10. "Moving on from Picasa," Google Picasa, accessed June 10, 2016, https://picasa.google.com/.

11. Daniela Hernandez, "The New Google Photos App Is Disturbingly Good at Data-Mining Your Photos," Fusion.com, June 4, 2015, http://fusion.net/story/142326/the-new-google-photos-app-is-disturbingly-good-at-data-mining-your-photos/.

12. Mike Ashenfelder, "The Library of Congress and Personal Digital Archiving," in *Personal Archiving: Preserving Our Digital Heritage*, edited by Donald T. Hawkins (Medford, NJ: Information Today, 2013), 31–45.

13. Denise Levenick, *How to Archive Family Photos: Step-by-step Guide to Organize and Share Your Photos Digitally*. Cincinnati: Family Tree Books, 2015.

14. Oliver Slattery, Richard Lu, Jian Zheng, Fred Byers, and Xiao Tang, "Stability Comparison of Recordable Optical Discs—A Study of Error Rates in Harsh Conditions," *Journal of Research of the National Institute of Standards and Technology* 109.5 (September 2004): 517–24.

15. Fred R. Byers, *Care and Handling of CDs and DVDs: A Guide for Librarians and Archivists* (Washington, DC: Council on Library and Information Resources and National Institute of Standards and Technology, 2003).

16. "SD FAQs," SD Association, accessed June 10, 2016, https://www.sdcard.org/consumers/faq/.

6

Other Media

Video, Audio, Genealogy, and Problem Files

This chapter looks at some common multimedia and special files that have not yet been covered, including audio, video, and GEDCOM genealogy files. All of these file formats have additional considerations beyond the regular file management described so far. Planning, creation, organization, and management of each file format will also be covered. This chapter will also discuss how to deal with old and problem files, such as those you can no longer open. These may include old messages from long-defunct email accounts, or Corel Draw files that won't work now that the software is defunct. In either case, the section on problem files will list what options exist for files that your computer can't read.

VIDEO

Certainly, video is related to digital photography, which is covered in the previous chapter. You will see many similarities. Just like with photos, people now create a lot of video on mobile devices. Also like photography, video was once a physical medium requiring film and could, at that point, be considered a fairly straightforward medium. Users with analog video, including film and VHS, can look to chapter 8 for information about digitizing. With the advent of new technologies and digital video, however, moving image has endured a "steady progression of new formats."[1] Because of the complexity of video file formats and the nature of digital video on the web, archiving digital video varies in difficulty depending on what you want to do with it. This chapter will cover the basics of digital video use, management, and archiving, and it should be enough for the casual computer user to understand and deal with digital video on their devices. Users with more advanced needs may need to seek out more in-depth resources specific to digital video management.

Video File Basics

Before getting started with digital video, it is important to understand how video file formats are set up. Each video file has up to three components: a container/file format, a video-coding format, and an audio-coding format. For example, a file with the MOV file extension (such as "dogskateboards.mov") is a QuickTime video file. The container is QuickTime, and it supports multiple video- and audio-coding formats. The file extension, MOV, will usually indicate what software or hardware is needed to open and play the file. An additional component called a codec is needed to decode/encode and compress/decompress the video file, which will allow you to open it. Wikipedia maintains a table of video file formats, their extensions, containers, and coding formats if you would like to get a broader picture of the digital video landscape.[2] In short, it is messy: there are a lot of options, and it is easy to get confused.

No video format is clearly the best. There are some common formats that work for many users, and basic video users should be able to use those common formats. On Apple devices, the default format for new video created on a device is MOV (QuickTime), which can be played on both Apple and PC machines. MOV files are often large because they are of higher quality. On other devices, MP4 (Moving Picture Expert Group 4) is a common format. These formats are playable on multiple platforms, including web sharing on mobile and desktop browsers. Both MOV and MPG/MP4 are decent options for general use. You should not have any trouble opening them on different computers, and both formats are in high enough use that they should not disappear in the near future—at least not without lots of tutorials for converting them to another option.

It may be helpful to know a few other video formats so that you understand your options if you do have the choice to import or convert video media. Also, like most people, you probably have a variety of older video files beyond those that you are currently producing, all in various formats. Here are some common video formats that you may own or come across, although this is far from a full list:

- AVI (Audio Video Interleave) is an older format that is at risk of obsolescence. You do not want to output new videos to this format. However, there are still readily available options for playing AVI files, so it is not a problem format quite yet.
- AVCHD (Advanced Video Coding, High Definition) is a format used primarily by Sony and Panasonic. You may have these files if you use these brands. This format is still in active use.
- WMV (Windows Media Video) files are common compressed video files often intended for web use due to smaller size. The trade-off for smaller size is reduced quality.
- FLV (Flash Video Format) is a primarily web-focused file format that is popular on online video-sharing websites like YouTube. Files using this format can be played in most Internet browsers as long as the browser has Adobe Flash Player

installed (users without Flash Player will be prompted that they do not have it installed when they try to watch a video online).

TIP: Unlike audio files, video formats are not usually available uncompressed; the common formats are compressed (some very efficiently to reduce quality loss) because uncompressed video files would be unreasonably large.

Video Storage

Size is a problem for digital video files. Videos get big fast, and that makes them difficult to store, as well as share. This encourages some users to consider more "lossy" compressed file formats, which create lower-quality (but smaller) files. If video quality is a concern for you, then you will purchase additional storage before choosing space over quality. If you are working with a lot of video, be aware of your computer's storage capacity and make sure that you are not getting close to filling your available space. You will also need to check your backup drives, including any cloud storage limits you have and external drive storage devices. To check your available space on an Apple computer, click on the apple in the upper left of your computer screen and select "About This Mac." Selecting the "Storage" option from the top of the window will allow you to see your current available space, as well as space on any external drives currently plugged in to the computer (see figure 6.1). On a

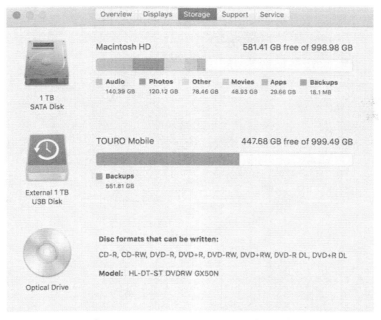

Figure 6.1.   Check computer storage space to make sure there is room for digitization projects.

PC, select "Computer" from the Start menu to see all current drives or to navigate to your C: drive. Your Local Disc (C:) will show a bar beneath it with the amount of free space. To see more detail, right click on the drive and select "Properties."

TIP: Because they are some of the largest files you are likely to own, weeding out video files is important to freeing up space. If you are low on storage space, reviewing videos for unneeded files is a good place to start.

Video Management and Organization

How you choose to organize your videos depends on your personal needs and the purpose of the videos. Because many people consider video to be a continuation of photography, you may choose to store and organize your home videos in the same locations and folder structure as photographs (see chapter 5). There are multiple advantages to this option. First, your videos are uploaded and transferred with your photos if you do phone-to-computer transfer or backup, and organizing them in the same way will save time. Second, it means that videos of your cat's birthday will be saved in the same location as the photos taken at the same time: you will not need to look in two places to find media from the same event. Alternatively, if you use video for other purposes, such as for art or work projects, then those video files may need to be stored with other related files. For example, a video inventory of your household for your insurance company should probably be stored with your insurance documents. It is okay to store video files in multiple locations if they serve different purposes; there is no reason to store these in the same location just because they have the same format.

TIP: Even if you have the technology to burn all of your movie collection DVDs to your computer, the size of the video files (while maintaining quality) is not worth the effort. Commercially produced movies on DVD are cheap to replace when compared to data storage.

As with other files, it is important to move your video files to your computer or other central location so they can be part of your computer backup. iCloud, Dropbox backup, and most other cloud storage applications will upload videos along with photos. The size of videos sometimes causes upload issues, however. If your video should be backing up to the cloud, check your account to make sure that large videos are transferring. If they are not, check with your service to see if there is a limit to the size of files that will be uploaded. Large files may also have problems transferring from devices over wifi, and you may need to open the upload application and keep it open until the entire file has been uploaded.

TIP: Remember that video files are large. Before transferring a large number of videos to your computer, make sure you have lots of extra storage space or purchase an external hard drive to compensate.

AUDIO

Audio is another popular digital medium, with a large portion of audio collections devoted to music. Even if you don't have a lot of music, you may have saved voicemails, audio notes or dictation, ringtones and sound effects, oral histories, or sound clips of friends. Some people have a lot of archival audio needs because they have family history projects, such as "letters" on cassette, recordings of meetings, or other events that were more likely to be audio recorded before video became easy to produce. In any of these cases, there are some basic things to know about audio files, their management, and their organization.

Audio in physical form is an at-risk media type. Casey and Gordon explain that "recording formats deteriorate over time, degrading much more rapidly than paper-based archival documents. Audio recordings rely upon reproduction technology that adds wear, fosters deterioration, and eventually becomes obsolete."[3] In other words, most media formats used for audio over the years are not great (think 8-track and cassette), and this puts audio at risk. If you have noncommercial audio on old media, whether the files are digitally readable or not, converting them to be stored digitally will protect them. Noncommercial is specified because the risk of losing most commercial audio (usually professionally produced music) is not as great, as it can be repurchased.

Audio File Formats

Like video, audio file formats can get a little complicated. However, there are clearer standards for audio than there are for video because the files no longer have the visual component (obviously, a video file must consider both sound and visuals, whereas audio only needs sound). Also like video, compression is a common issue for audio files. In general, choose uncompressed audio files for long-term preservation and consider lossy compressed formats for commercial and noncrucial audio files that you don't want to take up a lot of storage space.

Here are some of the most common audio file formats:[4]

- WAV (Wave Form Audio) is an uncompressed file format standard in archival audio storage.
- AIFF (Audio Interchange File Format) is an uncompressed file format native to Apple products but usable with many third-party applications.
- MP3 (MPEG-1 Audio Layer 3) may be the most common audio file type. It is a "lossy" compressed file, which means that it has lower sound quality but also a smaller file size.
- AAC/M4A (Advanced Audio Coding) is a lossy compressed file common on Apple products but playable on PC machines with some applications. This is the file format for most audio created on iPhone and other Apple devices.

- WMA (Windows Media Audio) is a lossy compressed file format common to PCs but playable on all computers with certain applications.

There are many other audio files that offer a variety of options and compatibility. In general, it is good to know whether the file format you choose is uncompressed or lossy. You may also see formats that are "lossless compression," which falls somewhere between the two, offering smaller files than uncompressed types and moderate quality options. It is also important to note that converting lossy audio files to uncompressed files (such as MP3 to WAV) will not result in better audio, only larger file sizes. Similarly, converting an uncompressed file to a lossy format will reduce the quality permanently; once an audio file is saved as a lower-quality file, any previous file data cannot be recouped.

If you are creating new audio or transferring audio from CD or other media, choose uncompressed formats for important or archival uses. WAV and AIFF are both uncompressed file formats suitable for high-quality recordings and archiving.[5] WAV is owned by Microsoft and therefore more common on PC computers, whereas AIFF is native to Apple. However, applications available to both operating systems can work with these file formats.[6]

Audio files of already low quality or of limited importance (such as ringtones, some music, and sound effects) may be better as MP3 files, or similar files of smaller size. Commercially produced audio (like music CDs) can be ripped easily and compressed to take up less space. This is covered in the next section. However, there will be some quality lost unless you opt for noncompressed audio, which takes up a lot of storage space. The audio on professional music CDs is uncompressed and of high quality to begin with, so ripping files at that higher level will retain better sound.[7] Whether this is important depends on the individual, as many people do not recognize the difference in quality.

Audio File Management

You probably already have a music management software program on your computer. iTunes is the most common, but there are many other free options. Even without the management software, your computer probably came with additional programs that will open sound files. Windows Media Player, QuickTime Player, and VLC Player are common audio programs. In addition, Audacity is a free program widely used by amateurs for modifying audio files. If you use a music program like iTunes, you may find that double-clicking on any audio file will bring up the program by default. This can be an issue if you don't want audio dictation, ringtones, voicemails, and oral histories in your music library. When opening those files, right click and choose Open With, then select a different audio program, such as Audacity. This will allow you to listen to the audio without adding it to your music library automatically.

Whether you have created audio on a device or have it on other media, transferring it to your hub location will allow you to modify, organize, and back it up.

Transferring files from audio CD is relatively simple if you have a CD drive on your computer. If you do not, you can buy an external drive that will plug into a USB drive for about $40. On many computers, simply putting in a CD will bring up a window asking what steps the computer should take next. Choose to copy or save the files from the disc to a specified location: your music library for music files, or another location for audio that you do not want as part of your music collection. There are also applications that can assist you with creating audio, and most have built-in options for backing up files to a computer or cloud storage option. In addition, any audio captured or playable on a cell phone or other mobile device can be transferred using Audacity, in the same way voicemail archiving is described in chapter 4. Transfer of old tape media, like audio cassettes, is discussed in chapter 8 because that audio is not yet digital and requires additional steps.

TIP: If you use a music organization program, make sure you know where your music files are stored on your computer. If you use iTunes (the most common), you can find your music files by going to your hard drive > Users folder > Music folder > iTunes. Music files may be in an iTunes Music or iTunes Media folder, organized by musician name.

GENEALOGY FILES

Genealogical records have their own file type specific to the needs of family historians and the systems they use. Genealogy files from many different software programs and online services can be exported as GEDCOM (Genealogical Data Communications) files.[8] GEDCOM files have a GED file extension, for example, "sherosky.ged." The history of the GEDCOM format begins with the Church of Jesus Christ of Latter-Day Saints, an organization known for its interest in family history. According to Mike Ashenfelder, the LDS church has been "gathering genealogical information since 1874, microfilming international family history records since 1938 and digitizing the microfilmed records since 1999. . . . In 1984, the LDS church developed the GEDCOM (pronounced "jed-com") specification as a means of exchanging genealogical data."[9] As one of the largest and most organized groups doing genealogical research, the LDS church's GEDCOM standard was adopted across all major platforms. You can certainly keep your family history information in many places, including text documents, databases, or even spreadsheets. However, since many genealogists use online services like Ancestry.com and Familysearch.org, GEDCOM files can make sharing and interoperability easier.

GEDCOM is text-based, but has a relationship-based data structure to allow for linking within records. Because it is text-based, file size is small and GEDCOM files can be easily shared via upload/download and email attachment.[10] If you use an online site or proprietary software to work on your genealogy, consider exporting a GEDCOM file periodically in order to protect your research. Include the date in the file name and store it on your computer so that you are protected if the system

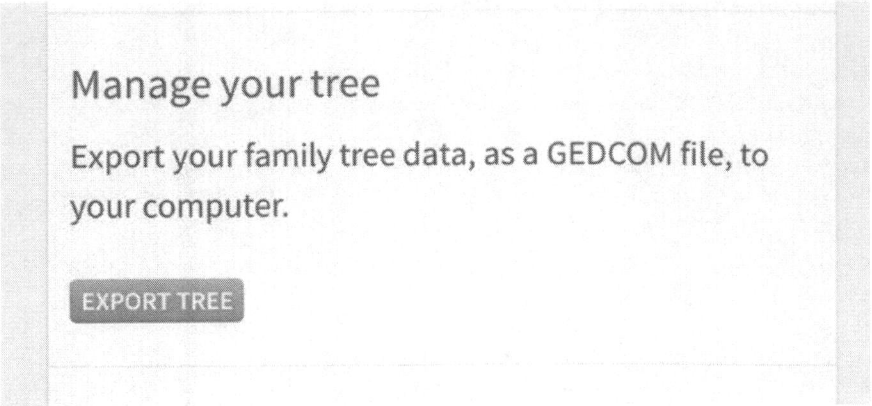

Manage your tree

Export your family tree data, as a GEDCOM file, to your computer.

EXPORT TREE

Figure 6.2.   Export your genealogy files from Ancestry.com using GEDCOM.

you use becomes unavailable. Don't trust that a genealogy site will hold your files and documents indefinitely, no matter how long it has been in existence.[11] This is especially important for family researchers who have devoted large amounts of time and energy to their work.

Exporting a GEDCOM file is different in each system, but in general you should look in the menu options for Export or Save GEDCOM options. On Ancestry.com, the export option can be reached by going to the Menu Bar/Name Family Tree (in upper left) > Tree Settings > Export Tree button on right of screen (see figure 6.2).

One of the only drawbacks to GEDCOM is that, although it is an open format, most computers do not have a tool automatically built in to recognize and open the files. However, even if you do not keep a program on your computer to open the file, you will be able to save and upload the file to other genealogy websites or download a new genealogy software program to see the file. If you mainly use an online tool for genealogy, then saving a backup file (even if you cannot open it at the moment) still protects you from loss.

PROBLEM FILES: HOW TO ACCESS
AND CONVERT ITEMS THAT WON'T OPEN

Nothing drives home the importance of using open file formats more than trying to open old files that can no longer be recognized by your computer. Unfortunately, quite a few files were created in old, proprietary files before many computer users recognized the importance of using open formats. Many new applications still use proprietary formats because they want you to continue using only their programs. Luckily, even when that is the case, newer programs often have an option to export or save as an open format. That means that, hopefully, you have a limited number of problem files that you cannot open.

The first thing to know is that you should not throw away a file just because you cannot open it. This is like throwing out a can of food because you cannot find the can opener. The file (like the food) is still good—you just need to find the right tool to open it. The good news is that tools are improving and becoming more common. Computers are in such wide use that many previous "problem" file types are now accessible because someone has built a converter or developed a solution for the problem. Even if you cannot find a fix right away, do not lose hope: hold on to a file unless you are running out of room or know that you no longer need it.

Here are the basic steps for dealing with old or problem formats:

1. Identify whether the problem item is personal or commercial/public information. If the file is an old video game, a really old report from the Montana Department of Motor Vehicles, or a CD-ROM of all of the issues of Mother Earth News from 1980–1989, then you should consider whether you want to spend time and effort breaking into the file. Instead, see if you can find a current source for the item. Try the Internet Archive (Archive. org) for old video games and websites (see figure 6.3). Try hathitrust.org for old public documents, and your local library for magazine articles and other publications.

2. Once identified, prioritize personal items over commercially recorded media. Commercial media (music, video, images) can be replaced in a worst-case scenario by repurchasing items, whereas your personal media usually cannot.[12]

3. Identify the file type from the file extension, which is the code after the period in the file name. For example, addit477.lwp can be identified as a Lotus Word-Pro file because that is the only file type that uses the .lwp extension. Wikipedia offers a constantly updated list of file formats that will help you identify a file type from its extension (just search for "Wikipedia list of file formats").[13]

4. After identifying the file type, consider whether you have the program needed on another device. For example, you may be able to open old Adobe InDesign files on your old laptop since the old version of InDesign is still installed there. If that is the case, save the file to a flash drive and open it on the other device. Once open, check to see what export or save options exist—you may be able to save the file in a better format (such as PDF). You can always keep the legacy file, too, if you are concerned about loss between formats.

5. If you do not have access to a program, look for a conversion tool that works with your file type. Numerous free online converters (fileminx.com, zazmar .com, online-convert.com) are available with varying results. Formatting of some documents may be changed during conversion. You can also purchase file conversion tools, as many are available—again, check to make sure they work with the file type you need to change. Do not download a free tool unless you can establish that it is reputable (reputable tools may have lots of positive

reviews online or be recommended by well-known technology websites like Wired or PC World).

6. If you cannot find a converter that will work with your file, consider buying or trying to find a copy of the old software. This may depend on whether the program will still run on your computer, so look at specifications and check before spending money.

7. If you cannot find a way to convert the file yourself, look for a service that will convert old formats to newer formats. Services like Retro Floppy (http://www .retrofloppy.com/) not only convert old media, but will also convert files. Keep in mind that you will be giving the service access to your file, which can be frightening if you do not remember what is in the file.

8. If no other option has worked, consider holding on to the file and look for a solution in the future: new conversion tools and applications are created all the time, and it is unlikely that you are the only person to have an old file in a particular format.

Figure 6.3. Internet Archive hosts millions of files and allows users to search archived websites through history using the Wayback Machine.

AVOIDING FUTURE FILE PROBLEMS

After dealing with problem files, it is easy to think about ways to improve your file management to avoid file issues. Here are some tips for avoiding file problems in the future:

- Use open file formats whenever possible.
- It may make sense to keep a file in multiple formats. For example, resumes are often created in Word and exported or saved as PDF files. Keeping the PDF versions will allow you to see the exact document sent to others, while keeping the Word document allows you to retain the editing options. Similarly, graphic designers might find it helpful to keep the proprietary editing documents while also keeping the EPS or image file of the final projects sent to clients.
- When using proprietary software, choose well-known names that have been around for a while. These companies are more likely to offer tools for transferring your stuff if their product changes, and other users are more likely to create tutorials on how to save or convert things.
- Name things well for easy identification. If a file cannot be opened later, it is very helpful to know what it is so that you know whether it is worth the time to investigate.
- While file corruption does not happen often, it does happen. Sometimes files stop working. Having backups is the best protection against loss from file corruption.

SUMMARY

After reading this chapter, you should know the basics of video, audio, genealogy, and problem file management. Combined with earlier chapters, the information here should round out the management of digital material stored on your computer and devices. The next two chapters will look at online accounts and materials that still need to be digitized, respectively.

NOTES

1. Richard Rinehart and Jon Ippolito, *Re-Collection: Art, New Media, and Social Memory* (Cambridge, MA: MIT Press, 2014).

2. "Video File Formats," *Wikipedia*, accessed July 16, 2016, https://en.wikipedia.org/wiki/Video_file_format.

3. Mike Casey and Bruce Gordon, *Sound Directions: Best Practices for Audio Preservation*, 2007, http://www.dlib.indiana.edu/projects/sounddirections/papersPresent/index.shtml.

4. Aimee Baldridge, *Organize Your Digital Life: How to Store Your Photographs, Music, Videos, and Personal Documents in a Digital World* (Washington, DC: National Geographic, 2009).

5. Dan Daley, "Codec Confusion: Deciphering the Alphabet Soup of Digital Audio Format Acronyms," *Residential Systems* 13.4 (April 2001): 20–23.

6. Casey and Gordon, *Sound Directions*.

7. Daley, "Codec Confusion."

8. "GEDCOM," *Family Search Wiki*, accessed July 15, 2016, https://familysearch.org/wiki/en/GEDCOM.

9. Mike Ashenfelder, "Family History and Digital Preservation," *Perspectives on Personal Digital Archiving*, March 2013, n.p., http://www.digitalpreservation.gov/documents/ebook-pdf_march18.pdf.

10. "GEDCOM," *Family Search Wiki*.

11. Ashenfelder, "Family History and Digital Preservation."

12. Baldridge, *Organize Your Digital Life*.

13. "List of File Formats," *Wikipedia*, accessed August 1, 2016, https://en.wikipedia.org/wiki/List_of_file_formats.

7

Social Media, Online Sharing, and Online Accounts

This chapter covers some of the things in your digital life that you may feel like you have little control over, in part because they don't live on your computer. Social media sites—including Facebook, Twitter, and even the professional networking site LinkedIn—can be archived in a number of ways. In a less public vein, this chapter will also look at online sharing of files, folders, and documents. This section includes checklists for how to handle collaborative documents and files, as well as the risks and benefits of sharing access. Finally, this chapter will also address managing online accounts. This will include keeping track of your online accounts and passwords and tips for effective creation and use of passwords.

SOCIAL MEDIA

Chances are, you have at least one social media account. Social media is a term used to describe interactive websites where people take part in social activities, including photo and video sharing, networking, information sharing, online chat, and personal (or public) status updates. According to Facebook, on average 1.13 *billion* users log in to the social media site every day—and Facebook is just one of many social media offerings.[1] Social networks have sprung up to cover almost all activities, from recipe sharing (Recipefy) to coordinating local groups on all interests (Meetup). Users may feel like Facebook and other platforms will never go away. Certainly some will have more staying power than others. However, the safety of your media and communications on the site can never be as safe as things that you have control over and back up.

Though we are faced with the constantly changing nature of the Internet every day, somehow many users of social media believe that it is a trustworthy holder of their information. When people add thousands of photos on Facebook and point

111

out that the platform has been around for years, this gives a false sense of security. If your only copy of a photo is on Facebook, then it is probably not backed up, and that means it is vulnerable. Facebook may be around for many years, but the platform may change and evolve like it has already, and just like other social media platforms do. Take Myspace as an example: one of the first social media platforms, and an extremely popular one, it nevertheless declined in popularity after many years of success. Very few people spent time going back to their account to see if there were comments or photos they wanted to save—they just stopped using the site. In a *Businessweek* article, "The Rise and Inglorious Fall of Myspace," Felix Gillette writes, "One of the reasons social networks are so combustible is that they have proven to be particularly sensitive to public perception."[2] So the question, then, is how much do you trust the public? Whatever social networks you use, you are likely to start paying less attention to them when something more popular comes around. By the time you remember that you left all of your graduation photos and well-wishes on Friendster, it may be too late.

Again, there are social media sites for almost everything. Some will come and go before you know they existed, particularly if they have a small market. This chapter will cover archiving options for the largest social media sites of 2015: Facebook, Twitter, LinkedIn, Instagram, and Pinterest (see figure 7.1).[3] If you participate on other social media websites—especially small or narrowly focused ones—be aware of anything posted there that you may want to save for later. Look for archiving or export options, and use them if they exist. If they don't, consider anything posted

Figure 7.1.   The most popular social media websites for American users

to the site as fleeting, unless it is posted publicly. If your social media site is public, then it may get grabbed by Internet Archive. For example, Recipefy's public-facing pages are being saved for all time, comments and all.

Facebook

Facebook is by far the most popular social network, so it is not surprising that it has some built-in tools for archiving your account data. It is relatively easy to save much of the information in your account, and the tools have been improved to capture much of the social interaction that makes social networks so appealing.

To create your archive,

1. Open Facebook on a computer.
2. Use the down arrow in the upper right-hand corner to view menu options.
3. Choose Settings.
4. On the new screen, select "Download a copy of your Facebook data" from the bottom of the General Account Settings list (see figure 7.2).
5. On the new page, click on "Start My Archive."
6. You will need to confirm your password, and the page will let you know that the archive may take a while. Depending on how long you have been a Facebook user, it could be a lot of data!
7. Once the archive is finished, you will receive an email at the address in your account, and you must type in your password again to download the files. The link in your email will expire after a few days. These are security measures to protect you from anyone else downloading all of your personal information.

The downloaded ZIP file will contain folders for HTML, photos, and videos, as well as an HTML index page (see figure 7.3). HTML pages can be viewed with any

General Account Settings

Name	Melody Condron	Edit
Username	http://www.facebook.com/MTbeekeeper	Edit
Contact	Primary: melody.condron@gmail.com	Edit
Password	Password last changed over a year ago.	Edit
Networks	No networks.	Edit
Temperature	Fahrenheit	Edit

Download a copy of your Facebook data.

Figure 7.2.   It can be easy to miss the download option at the bottom of Facebook's General Account Settings.

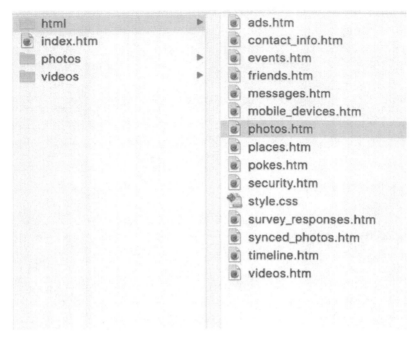

Figure 7.3. A variety of files will be available in a Facebook archive when it is unzipped.

web browser. The archived posts and comments have come a long way from previous versions of Facebook's archive feature, and you can see comments on groups of photos as well as individual photos in a group. Photographs in the photos file, however, are more difficult to manage. They will all be JPG files with long numeric names—not exactly helpful for archiving if you have hundreds of photos. Beyond generic names, all photos have been stripped down in size (72 dots per inch [dpi]), any location or keyword metadata has been removed, and unintelligible coding has been added to the IPTC metadata as "instructions" (see figures 7.4 and 7.5). While it is great to be able to log the photos posted and download a copy if all others are lost, a Facebook archive is not a great way to save your photos.

Facebook is also very specific about what data is included in your archive. If you are interested in seeing the long list of what is included, hit "And more" from under "What's included?" to see the Facebook list of "What categories of my Facebook data are available to me?"[4] Remember, privacy and settings change often on social networks, so this list may change. For example, Facial Recognition Data and Posts to Others were not on this list in 2012, and you find many of the items on the list have also changed.[5]

Facebook is the only major social media site (so far) that additionally allows users to set up legacy controls. That means that you can decide what will happen to your

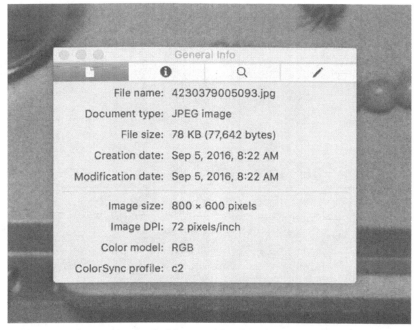

Figure 7.4. Photos downloaded from Facebook in the archive are of poor quality.

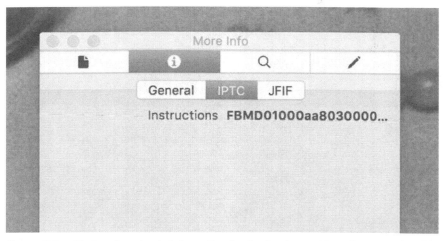

Figure 7.5. Photos downloaded from Facebook also have stripped or questionable metadata.

account after your death. You have options to continue the page as a memorial, with a Facebook contact set up as the facilitator. That person will be able to download your Facebook archive just like you can. Alternatively, you can have the page set to shut down if someone alerts Facebook that you have passed on. Legacy controls can be found by clicking on the down arrow in the upper-right corner of the screen and selecting Settings > Security > Legacy Contact. There you can choose a legacy contact, give them archive options, or opt for account deletion.

Twitter

Twitter is a public-facing social media site, which means that anything posted there can be archived not only by you, but by anyone else who wants to take the time to do it. The Library of Congress, for example, started archiving all tweets in 2010 (though it has yet to make them accessible or usable in any meaningful way).[6] Depending on how you use the platform, you may want to create an archive of your posts, especially since there is a built-in tool for users.

To create an archive of your Twitter account posts, go to your Profile and Settings options by clicking on your image icon in the upper right, next to the Search Twitter box and Tweet icon. Under Content, select the button for "Request your archive." An archive of all of your tweets will be compiled and a link will be emailed to you at the email address listed in your account (see figure 7.6). Clicking on the link in your email will allow you to download your archive as a ZIP file. Links to the archive do expire, so make sure to download the file soon after receiving the link. The ZIP file includes a CSV file that can be opened in an Excel or other spreadsheet program, allowing you to see all of your tweets, retweets, and links, each marked with time and even posting device. The archive also includes an HTML file viewable in a web browser, showing not only a better view of your tweets (including images) but also a yearly bar graph of activity (see figure 7.7).

What is missing from the archive is the interaction. Unless you retweet someone's responses to your message, they are not included in your archive. "Hearts" and num-

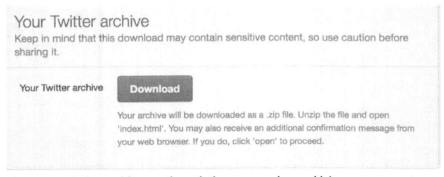

Figure 7.6. Twitter guides you through the process when archiving your account.

Figure 7.7.   A downloaded Twitter archive comes with a graph of our personal user activity.

ber of retweets are also not included, nor are your replies to other people's tweets. Twitter's public platform should mean that privacy issues are not part of the problem. Instead, it is likely that the large amount of data connections and interaction make archiving the connections difficult.

LinkedIn

LinkedIn is a professional social media site used for networking and job searching. Online profiles on the platform consist of job history and professional skills, though users can also participate in discussions and groups and add status posts. There are a number of reasons why you might want to download your LinkedIn account information. Primarily, since it is a current list of your professional activities, you can use the information on the pages as the starting point for a new resume. If you keep LinkedIn updated, it may be a more accurate starting point than old resumes. As with keeping older resumes, you may find it helpful to have a running list of your previous positions and involvement. You can also download a list of contacts, which can be helpful as a backup or for loading into another contacts program.

Archiving your main profile page is remarkably simple. Navigate to the page and select the down arrow to the right of the View Profile As button, just below your photo. Choose Save to PDF to save your profile page information, which can be easily adapted to a resume. To download your LinkedIn contacts with names, email addresses, employers, and positions, mouse over My Network in the top bar and choose Connections from the menu. Click the Setting icon (a gear) near the top right, then select Export LinkedIn Connections in the right bar (see figure 7.8). Contacts can be exported as a CSV or VCF file for access in spreadsheet, database, or address book applications.

Instagram

Instagram is a popular photo-sharing site that requires an application on your mobile device in order to post. While you can log in to your account from a computer, interaction on the platform it set up to be primarily mobile, and account management settings are limited. The site has no export or archive options built in. Photos are uploaded/imported from your device to the site, and that may be fine with some users. However, the filters applied to each image are only on Instagram and cannot easily be saved due to features designed to stop unintentional copying. While there are a number of unauthorized ways to capture photos even without built-in tools, a third-party app called InstaPort (https://instaport.me/) has filled the need for Instagram archiving. Entering your user name at InstaPort will allow you to select which photos you want to archive and download them as a ZIP file. Photos in the ZIP file are in JPG format. Not only can you archive your own photos this way, but you can

Figure 7.8.   The Export LinkedIn Connections is not easy to see.

also create an archive of other users' photos, assuming they are public or have set their privacy settings to allow it.

Pinterest

Pinterest is possibly the most difficult social website to capture long term. The site pulls images and information from multiple sources to create "boards," which is its strength. However, that means that data is not easily capturable from the aggregate page that users see tiled in front of them. Many users have been successful using the "Print to PDF" option available under the File menu. However, this often cuts off posts or the sides of pages due to how the site is formatted using tiles. Capturing the pages in HTML is a decent option, though it does mean that some images may stop showing over time as links die. However, some users may like to have an HTML snapshot, which will not change even if the board is changed and can be accessed even if the user isn't logged in to the site. To capture a board or other page, first go to the page you want to save using the Mozilla Firefox web browser. In the sidebar next to the image posts, right-click and choose Save Page As. . . . In the pop-up window, adjust the file name as needed and select Web Page, Complete from the drop-down menu (see figure 7.9). Though the file will open in a web browser and look just like the website, you can see in the address bar that you are no longer seeing a hosted page on the Pinterest website.

Archiving the Web in General

Here are some options for saving or capturing an individual web page:

- First, check the Internet Archive Wayback Machine at archive.org to see if the page you want is being archived. If it is, you will be able to see not only the current page, but multiple versions of the same web address as it changes over time.

Figure 7.9. Pinterest, which has no built-in archiving function, must be archived using tools that are not ideal.

- Download the HTML code used to build the webpage, which stores the data needed to rebuild the page. To capture a page in this way, go to the page you want to save using the Mozilla Firefox web browser. While mousing over the page (but not a specific image), right-click and choose Save Page As . . ., then in the pop-up window, add a file name and select "Web Page, Complete" from the format drop-down menu. This will save the webpage as an HTML document that can be opened later in any web browser. Unfortunately, since the original HTML may rely on links and online images, some of the functionality from the original page may not exist when the archived file is opened.
- Most computers offer the option of saving a file as a PDF. With the webpage open in a web browser, select File from the options bar, then Print. Look for the Print to PDF or Save as PDF options. Saving as a PDF will often allow long webpages to be saved as multipage PDF documents complete with images. Because the web format does not always translate to the standard-sized PDF pages, formatting may be different than how it appears on the webpage.
- If you want an exact copy of the image you are seeing on a webpage, you can always take a screenshot (a photo capture of your screen). On Apple, select Command + Shift + 4 and crop the area of the screen you would like to capture, then find the saved PNG image file on your desktop. On a Windows 8 or 10 PC, hit the Windows Key + Print Screen ("Prt Sc") button in the upper-right corner of the keyboard to capture the entire screen as a PNG file. The file will be automatically saved in a Screenshots folder in your Pictures directory. If you are using another version of Windows or just select Print Screen, the image will be saved to your clipboard and you can open any program to paste the image and save.

ONLINE SHARING

There are many reasons why you might want to share documents and files online. Making sure everyone has the latest version, allowing everyone to contribute to a document's creation, and ease of access in transferring large numbers of documents between people and computers are common reasons. Of course, you can always email a document or copy it to a flash drive to give to someone. However, emailing files that change regularly can be a hassle, and passing along a flash drive means you must be in the same physical location or mail the drive. Sharing online files makes access instantaneous and means that everyone is seeing the same thing. As discussed in chapter 3, family and loved ones often have a difficult time accessing files after someone's death, in part because the legality of that access is not clear in many states and countries.[7] One way around this is to share access to documents and files online. Couples and families, too, may find it easier to store files that relate to multiple people in one online location. Finally, sharing is also important for group projects, especially when contributors are not all in the same physical location.

Despite the many positive aspects of online sharing, there are aspects that make it more complicated. In cases of shared editing, all sharing partners can move, modify, and delete files. It can also be unclear whether online files are backed up and safe. If you already share online documents or plan to do so, it helps for everyone to be clear on the expectations and possibly set some ground rules, discussed later in this section. Questions to ask when sharing documents online, as well as basic information about setting up shared accounts and documents, are covered in this section.

Online Files: Setup and Sharing

Consider shared online folders and files like a joint bank account: if one person on a joint account is not available or (worst case) passes away, the other person on the account does not have as many problems accessing the account as if they were NOT listed as an account user. If nothing is wrong, the shared space is still easily usable by both parties, and people can both add things that the other can see and use. Shared folders and files work similarly. The one main difference is that shared folders and files do not usually mean that you have a shared account. Instead, both (or multiple) people are sharing a folder that can be accessed by multiple different user accounts.

There are multiple free online storage sites, and many of them allow folder and file sharing. Dropbox is possibly the most common program because it is free and has many easy-to-use capabilities. Some of the examples here use Dropbox, but many other online sharing systems have similar features. Other popular storage sites include Box.com, OneDrive, CloudMe, and Amazon Cloud Drive. Each service usually offers some space free, with additional storage space for a fee. Most online storage sites also have added encryption as a fee service if you would like to share documents containing personal or sensitive information.

Most online storage programs can be added to all of your devices and multiple computers so that you can access files from anywhere. You can also log in to your account online from the program website if you are using a computer that is not yours or does not have the program added. On your devices, the program will appear like any other app, and will allow you to store, save, and access files to the online location. On a computer, the program will look like a folder location and can be added to the locations in your computer sidebar (see figure 7.10). Most of the subfolders in your online storage account are probably not shared with anyone, and they won't be unless you choose to share them. In Dropbox, if you want to create a folder to share with others or share an existing folder with someone, right-click on the file, folder, or the blank area under your files (for a new folder). As shown in figure 7.11, you can either Share (which sends sharing capabilities via email) or select Copy Dropbox Link. If you copy a link and send it to a person or a group, these users will have access to the folder or file. Whether you email or share a link, you will have the option to allow the other participants to edit or to simply view the files.

There is a difference between sharing access to files and sharing documents. Once you give someone editing capabilities, they can change, move, and even delete a file.

Figure 7.10. Once loaded on a computer, Dropbox will appear along with other computer locations.

Figure 7.11. Right-clicking on a file or in a Dropbox folder will bring up options to share directly or with a link.

If your intention is to give viewing access only, you can do that without any risk of files getting changed or moved. If you use an encrypted storage option, papers like wills, insurance documents, and family histories can be easily made available to family members in this way.

When sharing online documents with editing capabilities, there can be more risk of loss and confusion. Here are some guidelines for best use of shared storage:

- Are any users concerned about storage space? If so, let other users know that you only want to use the shared folder for specific projects.
- Are the files in the online drive copies? In general, avoid putting your only copy of something in a shared folder that allows someone to edit or delete it.
- If you add files so that someone can have a copy to move to their own computer, make that clear so that that person knows he or she can take things out. Otherwise things sit in the folder and use up space. This is a common problem with photo sharing. If you regularly use online storage to send groups of items back and forth, consider having folders that indicate things going to and from each person, such as "Makayla to Arturo" and "Arturo to Makayla." This will help to keep things straight and unmixed in either direction.

Communal Work/Documents

If you are sharing files online in order to collaborate on a project or document, this can also be confusing. When working on collaborative documents, such as a shared Google Drive page or Word document in Dropbox, it is helpful to discuss expectations. Here are some basic issues to discuss with your collaborators to make sure everyone is on the same page:

- Does the document title work for everyone? If four people are collaborating on a Library of Congress pilot project and they call it "LOCproject," that may work for the three people who do not work at the LOC, but not for the one who does.
- Does the group prefer tracked changes, comments, or both? Or may all users make changes directly in the document without noting them?
- Will anyone keep a backup in a secondary location? If so, who, and how often will they back up the file? No one expects Google to go under any time soon, but other online storage sites might, and almost all sites have down time, even Google Drive.[8]
- Who will remove users if someone drops out of the group? It is not uncommon for ex-collaborators to have access to shared documents long after they have left a project, which is not always safe or desirable.

ONLINE ACCOUNT MANAGEMENT

A common complaint from computer users is having to keep track of all of their accounts, with logins and passwords. There is no easy solution to this, though many users find that password programs simplify their lives. For many people, just keeping track of where they have accounts and login information is helpful. Here are some basic tips:

- While it is not recommended to keep a list of your accounts with login and password, it is not a bad idea to keep a basic list of accounts. It is still a good idea to password-protect the list (described later in this section and in chapter 3).
- Share your list (again, without passwords) with a family member so they know what accounts to shut down or follow up with in the event of an emergency.
- Periodically weed your online accounts. If you keep a list, consider going through it for sites you no longer use. Log in to the site and delete your account.
- If you do not keep a list, you can sometimes use email as a prompt to weeding accounts. You will often get emails from sites where you have created an account. When this happens, you can unsubscribe from the bottom of the email and also log in and delete your account.

Overall, the trick with online accounts is to close the accounts you do not use, and keep track of the accounts you use regularly.

Password Management

There are hundreds of articles and blogs written about passwords. Every year, password statistics indicate that many computer users continue to choose poor passwords that do not protect them. Many people do not take password security seriously, especially once they have to manage 500 different account login combinations. However, passwords for banking, email accounts, and medical accounts should be given more thought and consideration. Social media password breaches can also become a very public problem if you are targeted. SplashData's 2015 list of the "worst" (most common) passwords shows that people continue to choose easy-to-crack words, numbers, and combinations to protect their accounts. The most common include 123456, password, baseball, football, abc123, trustno1, and letmein.[9] Knowing that these stay the same from year to year, hackers can easily use this list as a starting point when trying to break into your accounts. That means that using any of the passwords on SplashData's list makes you an easy target.

Still not convinced? Consider strong passwords to be like car seatbelts: you may go your entire life without needing one, but if you need it, you will be grateful you thought to use it. Seatbelts and strong passwords both prevent harm in worst-case scenarios. Many people wear their seatbelt consistently because they have developed a good habit in order to protect themselves. Consider doing the same with passwords and, though you will hopefully never need to rely on it, you will be protected if you or your data is at risk.

A Note on Password Sharing

Many password articles advise never writing down your password and never sharing it. As mentioned in the legacy planning section in chapter 3, this is not a practical expectation when considering emergency access for family. It is also incredibly unrealistic: many people will not commit to a password management program, and yet cannot remember their million passwords. If you do choose to share your passwords, do so with caution and choose your methods wisely. Consider, too, that the best choice may be to keep an offline list of passwords in a personal safe or locked filing cabinet, where it cannot be reached by remote computer hackers. Since much of your financial and personal information may be on printed documents in the same locations, your passwords should be relatively safe.

Password Tips and Suggestions

While the Internet is rife with suggestions for how to create great passwords, here are some methods that are relatively easy:

1. Use the beginning letter of every word in a phrase, with at least one number and special character thrown in as part of a word. For example the phrase "Man, I hate coming up with new passwords all the time" might become *MIh8cuw/npatt.*
2. You can develop a code that can work for multiple passwords on multiple sites. For example, if you have a core password that has a number, letter, and special character in it, you can then add something site-specific to that core for each website where you need a password. Adding the first few letters can work. For example, if your core is 2!7uP, you could choose to add FACE to it for your Facebook password, making it FACE2!7uP.
3. When changing passwords, consider changing the numbers in the passwords rather than the whole word. If you have become used to Jg7wpol%l and have it memorized, consider changing the 7 to an 8 when it is time to update.

Overall, changing your password often and using strong, uncommon combinations is the best protection. While you can be unceasingly vigilant about password protection, consider these basic password guidelines:

- Strongly consider using a password management program. These programs create strong passwords for all of your accounts and require you to only remember one (hopefully extremely secure) master password. PC Magazine ranks Dashlane, LastPass, and Sticky Password as the best options in 2016.[10]

- Special care should be given to your email account and any accounts that contain financial and medical data. If you cannot commit to strengthening and regularly updating passwords for all of your accounts, identify your most important accounts and manage those better.
- Do not use the same passwords for your important accounts, especially email accounts. Consider: if another site locks you out, it will often send you an email confirmation. If the password on your email is the same as the password on the hacked account, you are asking for trouble as both may be compromised.
- Do not email passwords to other people, especially for important accounts. If you need to share passwords for some reason, do it "manually"—in person.
- Do not, under any circumstances, create a document called "PASSWORDS" and store it on your computer. If you are hacked that list will work as a checklist for how to steal your entire identity.
- If you do create or keep documents with login and password information on your computer, for whatever reason, give them titles that do not imply that they contain secure information, and add password protection and encryption, described in this section.

When you need to create a document that does include password information or, for that matter, any personal, financial, or medical details, password-protect and encrypt it. To password-protect and encrypt any Microsoft Office document (Word, Excel, or PowerPoint) on a PC, click on the Office Button, then Prepare, and Encrypt Document. On a Mac, click on Preferences and choose Security from Personal Settings. Additional details and screenshots of password protection and encryption are available in chapter 3.

SOCIAL AND WEB ACTIVITY SUMMARY

After reading this chapter, you should be able to:

- Archive posts from social media accounts, including Facebook and Twitter.
- Save individual web pages as HTML, PDF, and screenshot images.
- Share documents efficiently by discussing common problem issues with collaborators.
- Keep track of your online accounts and passwords.
- Create better passwords.

NOTES

1. "Company Info," Facebook Newsroom, accessed August 31, 2016, http://newsroom .fb.com/company-info/.

2. Felix Gillette, "The Rise and Inglorious Fall of Myspace," *Bloomberg Businessweek*, June 22, 2011, 52–57.

3. "What Are the Top Social Networking Sites?" eBizMBA Website Rankings, accessed September 2015, http://www.ebizmba.com/.

4. "Accessing Your Facebook Data," Facebook Help Center, accessed September 1, 2016, https://www.facebook.com/help/405183566203254.

5. "Accessing Your Facebook Data," Internet Archive Wayback Machine from December 11, 2012, accessed September 1, 2016, https://web.archive.org/web/20121211045340/http://www.facebook.com/help/40518356620325.

6. Nancy Scola, "Library of Congress' Twitter Archive Is a Huge #FAIL." *Politico*, July 11, 2015, http://www.politico.com/story/2015/07/library-of-congress-twitter-archive-119698.html.

7. Jefferson Bailey, "The Big Digital Sleep," *Perspectives on Personal Digital Archiving*, March 2013, http://www.digitalpreservation.gov/documents/ebookpdf_march18.pdf.

8. "Google Drive [down time]," Down Detector, accessed August 26, 2016, http://downdetector.com/status/google-drive.

9. "SplashData's Worst Passwords of 2015," *SplashData* (blog), January 19, 2016, http://splashdata.com/blog/.

10. Neil J. Rubenking, "The Best Password Managers of 2016," *PC Magazine*, August 9, 2016, http://www.pcmag.com/article2/0,2817,2407168,00.asp.

8

Non-Digital Materials

This chapter will look at the non-digital: the stuff left behind, piled in boxes and in closets as we accumulate the digital stuff that already feels like too much to handle. There is light at the end of that tunnel. After many years of forgetting about some of that boxed-up paper, VHS tapes, and maybe even photo negatives, you may be better off. There are now more and (sometimes) better solutions for digitizing and preserving all of it. Digital storage and basic equipment are cheaper than they used to be. Books and the Internet are full of how-to instructions and tutorials to guide digitization projects. Businesses are available for nearly all old media if you don't think you can do it yourself. And many public items that you may have in your boxes may have already been digitized by libraries and archives! Though there is still some work to do, there are many options, and this chapter can help you get started.

First, it is important to note that there are many books related to digitizing, some specific to photos and others to audio and other media. This chapter can set the stage, but then you may need to go out there and get some more in-depth resources. Alternatively, after understanding the basics, you may feel like paying to have it all done for you is the way to go. Or it may be somewhere in between (digitize your own documents, send out the photos). No matter what you decide, you will need to follow the same steps, outlined here. Here are the basics of any digitization plan:

1. Make a List
2. Organize and Weed
3. Prioritize
4. Digitize

Before you start, make sure you have backups already running and have enough storage space available on your machines and backup drives/accounts.

STEP 1: MAKE A LIST

The first step in digitizing your physical items is to make a list. Instead of saying "all photos," list things in the units they are currently in: shoebox of photos, three albums of photos, two file-cabinet drawers of documents, and so on. This will allow you to tackle the project in chunks, such as finishing one box. That will give you clear stopping points while doing the project, since you probably cannot get it all done at one time. You may remember things later that are not on your original list, and you can add them at that time—just make sure you keep the list handy. Also, you can make a new, updated list any time you want.

While doing an inventory, "count, don't browse." In other words, don't get caught up looking at old memories and get off track.[1] List all the batches of items you would like to digitize, their format (DVD, letters, photos, etc.), importance level, and whether you will retain the physical items once they are digitized. See table 8.1 For an example of how to set up your inventory, or check appendix A for a blank inventory form you can copy and use.

Think about Originals

As you work on the next steps, be on the lookout for which things you will keep even if you digitize. For many items, you may find that you no longer want to keep the originals if you have a digital copy and multiple backups. Some items, however, may be important as physical objects even after a digital copy exists. These may include personal handwritten letters, original photos of personal value, artwork, and anything of high personal importance. In his chapter "Death by Technology," Ippolito offers, "When it comes to the causes of obsolescence, technology itself tops the list of the usual suspects."[2]

STEP 2: ORGANIZE AND WEED/SELECT

Organizing and choosing which things should be digitized may be the most important part of your project. It controls how much work there is, and also contributes to prioritization. Think closely about the reasons you have for digitizing. In many cases, preservation of materials is a form of memory assistance: we do not want to forget things, so we digitize and store them so we can refer to them later. It may be part of a project, such as a family genealogy or the creation of a digital scrapbook. In almost all cases, individual people who digitize and preserve materials (as opposed to institutions like libraries and archives) do so because they not only want to store and protect their materials, but also want to have better access to them: they want to be able to find and look at their records, photos, and emails later.[3] This strong connection between memory and preservation is an important part of the planning process.[4] If you want digital access and sharing capabilities for all of it, then digitize all of it!

Table 8.1. Inventory of Items to Digitize

Description (e.g., "photo shoebox #1")	Type (e.g., photos, VHS)	Volume (how many items)	Location (e.g., hall closet, attic)	Importance (1–5)	Keep after digitized? (Yes/No)

That will take a lot of time, but is worth it for some people. Yet if you are digitizing because you think that you should or you are supposed to, then you may want to do the minimum: the crucial documents identified in chapter 3, and maybe the best photos in your collection. No one says you *have* to digitize everything! Identifying motives will help you decide which items are a priority.

Consider the future use of digitized items, and also consider what will happen with all of your digital documents long after you are gone. Depending on our involvement, hobbies, family, and planning, you may want your items for personal use and expect them all to be deleted after your death. Conversely, you may find that you want your work and papers to be available forever.[5] This will also affect what you choose to digitize, and what you prioritize. Just remember, it is easier to find things when there are fewer items in a collection. This is true for you as well as anyone searching your collection in the future, so be selective.

Here are some suggestions for organizing and weeding before a digitization project:

- Sort documents into folders or piles based on what type of item they are.
- Sort and organize, then weed. If you think too hard about your items while you are sorting them, you may never finish. Instead, sort like items (medical documents together, negatives together, etc.) into groups and then assess the group. As an example, you may decide that you don't want to keep any bills older than one year, and it will be easier to weed those out after the bills are all in one pile.
- Sorting papers or photos before digitizing will also allow you to easily organize them digitally: you can scan them in batches and add them to a folder with the same name, or organize them with automatically generated subject names.
- Organize photos by date before digitization. It is OK if you don't know the exact date; start with year or even decade. Find some shoeboxes or large envelopes and start sorting photos into each date range. When you scan later, you will be able to add those date ranges to the folders and file names as they are scanned, which will save you the time it would take to label them manually.
- When considering whether to weed documents from your digitization plan, remember that you do not have to get rid of the paper copy: you just don't have to digitize it. Digitization should be for things you want to preserve, things you want to keep for a long time, or things you want to have digital access to for convenience. If you have other items that make more sense as physical items only, that is OK!
- Look at chapter 5 for specific suggestions on weeding photos.

After sorting and weeding, you may need to adjust your inventory, or cross things off the list completely.

STEP 3: PRIORITIZE

Once you have an updated inventory and a sorted, weeded collection of items, consider which batches of items you will work on first. If you marked things as high, medium, or low importance on your inventory, that should usually guide your approach. However, there are reasons to start in different ways, too. Consider these factors when prioritizing what to digitize:

- *Speed*: Do you have the know-how and equipment to easily scan all of your paper documents immediately? If so, getting them out of the way may be quick and a good way to clear space.
- *Space*: If you are low on space or trying to make room in storage, you might choose to prioritize items that you will no longer keep in physical form once they are digitized. CDs, for example, may be taking up a shelf or closet space that you could put to better use. Just remember to have backups set up, and check the digitized items before getting rid of originals.
- *Obsolescence*: Old cassette tapes or CDs, deteriorating papers or photos, or other items in rough shape might need to skip the line, or they won't make it to digitization.[6]
- *Impending Projects*: Are there photos and art that need to be digitized to meet a deadline, like a work project or a family reunion?

All of these factors can contribute, but in general you will look to your importance rating to help you. Digitizing those things that hold high importance first will mean that they should get done even if something comes up soon after to sidetrack your project.

STEP 4: DIGITIZE

When looking at a large group of items to digitize, you will need to consider how much time, money, and technological expertise you have to work with. If you have the money to spend and don't have the time or expertise, working with a business that digitizes and converts your material may be the best option. However, there is a risk of loss any time you have to entrust your material to someone else. There are many reputable businesses available, but accidents happen. If you must mail items to the business, then your materials are also vulnerable during the shipping process, and you sadly cannot insure photos for their personal value. Doing it yourself also has challenges and benefits. You can often borrow or rent equipment for a reasonable price, but doing it yourself will usually take a lot more time. Time involved will include the project time as well as learning any skills you may need for the project, including how equipment works. That being said, you will have much more control over the project if you do it yourself, and ultimately that is an important issue to some personal archivists.

Digitizing Documents

Documents are usually the easiest items to digitize. Printed pages of normal size can be scanned through a scanner document feeder, and do not have to be scanned at high resolution. Scanners with document feeders are relatively low cost (under $100), easy to find, and come standard on many "all-in-one" scanner/printer devices. You can scan a stack of documents all at once by adding them to the feeder; keep in mind that it may be best to do them in stacks of similar material so that you can scan them into one file (for example, you can scan a year of bank statements into one PDF document if they all scan together). Handwritten letters, papers having unusual sizes or shapes, or items in poor condition should be scanned using a flatbed scanner, where you place the item directly on the glass. Most scanners with feeders also have a flatbed, and some scanners only have flatbed. This takes more time since you must scan one item at a time. Scanning at 300 dots per inch (dpi) should be more than sufficient for most documents, and lower quality may be fine for unimportant documents with no images. If you do not have a good understanding of computers and scanners and want to have your documents scanned, you should have no trouble finding a local business that will assist you. Copy, office, and shipping stores often offer this service, and they can put the digital files on a flash drive for you to take home. Depending on their equipment and staff availability, some public libraries may also be able to help you scan documents, possibly for free or at a lower cost than other locations.

Before starting on document scanning, there are a few things to consider:

- You may be able to download many financial and medical documents from an online account, or have them emailed to you. This can save money and time.
- Fragile or unusually-shaped documents are better suited to a home project, especially if there are a lot of items in your collection; most businesses are not well-equipped to deal with unusual items.
- Using the services of a business or library does put you at some risk, and many people recommend using a private scanner to avoid unintentional copying of your documents. This is most important with sensitive documents that include account numbers and social security numbers.

Digitizing Photographs and Images

Dealing with a photo project can be daunting if your collection is large, as many are. The size of collections often make outsourcing the labor to a business unreasonably expensive; many photo scanning services charge $.04–$.06 per photo, or up to $1 per photo for hand-scanning on a flatbed scanner.[7] That can add up very quickly.

If you assess your collection and decide to use a service, there are many available. An Internet search for "photo scanning service" will help you find any in your local area, as well as the big, online companies that you can ship photos to in boxes. Most of these businesses specialize in photos, slides, and negatives, and sometimes VHS-to-DVD home-movie conversion. If you can find a business in your local area, this lessens the risk involved with mailing family photos. Using a scanning service involves more money but less time and no computer expertise. It does also involve some risk even if you use a local service. Most scanning businesses use auto-feed machines that are only suitable for photos of normal size, in good condition, with no tape or attachments. Otherwise, items could be caught in the scanner and ruined.

You will have more control over the project and protection of your photos if you do a photo-scanning project yourself. There are two main choices if you choose to scan items at home: purchase a home use scanner, normally for $100–$300; or rent a professional-grade scanner for a short time ($300 or more depending on how many days you need). Companies like EZ-Scan (www.ezphotoscan.com) rent the same high-end scanners that are used by photo-scanning companies. Rented scanners come with a computer and loaded scanning software and can scan up to 85 photos a minute. Both sides of the photo are scanned and you can batch-name each group of photos into one folder, depending on which scanner you use. Photos can be saved to a flash or external hard drive. At hundreds of dollars, professional scanners are expensive to rent (by comparison, the machines are several thousand dollars to purchase). If you do plan to rent equipment, block out time to do nothing else while you have the rental: get the entire collection scanned, and you may never have to rent or scan again.

If you choose to purchase a scanner (or borrow from a friend), read reviews online about the quality of the scans and any related image issues. Try to avoid using a scanner that gets poor reviews for photo scanning by other users. Choosing a scanner for photographs needs more consideration than one for documents, since range of accuracy and color capture are very important for these images. Test the scanner by using 300–400 dpi and adjusting settings. Look for sharpness, fine lines, and detail (all good signs), as well as "distortion, poor contrast between lines, false color, edge and halo artifacts around text."[8] Companies that sell scanners may advertise specifications that are not consistent, so the only true test is how your photos turn out. Check scanned photos as you do them, especially at the beginning of a project.

Here are considerations for photo scanning:

- To reduce effort and protect photos, keep photos in albums and do not attempt to physically repair them: digital tools can likely do a much better job.
- Check with your local library. Sometimes they are equipped to help people with small photo-scanning projects.
- If you have a small number of photos to scan, such as a handful of favorites that you want to prioritize, you may also be able to use one of the many scanning photo kiosks available at drugstores and other general stores. These kiosks are

not suitable for large projects, but you can easily scan a few photos to a flash drive for a reasonable price.

• Negatives are not photographs. Consider buying a lightbox for $20 to identify whether negatives are worth your time to digitize, or perhaps are duplicates of photographs that you still own. Unlike photographs, negatives can benefit from higher-resolution scanning: a minimum of 2000 dpi can help to capture the greater detail in the film image, and 3000 or 4000 dpi would not be overkill.[9]

As mentioned, there are numerous books available on digitizing collections. Nowhere is that more true than with photo digitization. If you choose to do a photo project on your own and have little to no experience with computers, you will want to seek out those books. Preparing in advance is the best way to proceed, and will prevent you from having to scan photos a second time.

Digitizing Video

If you have old video media, you may be able to transfer the video to your computer yourself if you still have the equipment. Otherwise you will need to acquire the old equipment or work with a business that will convert things for you. If you plan to transfer VHS or other analog video media to your computer, you will need the following things: an old player for the media you have (such as a VHS player); a computer with some sort of video capture software installed; an analog-to-digital converter (costs about $30); and cables for connecting everything. There are numerous video and website tutorials available for converting VHS tapes, which is a multiple-step process. Search the Internet for "how to convert VHS to digital" to find both video and written steps for conversion. You will notice that there are multiple options for tools, including software. Digital media, like early digital storage cassettes from the first digital video cameras, will follow a similar process, but without the converter. If you have video on DVD or another digital storage media, you should not have trouble transferring that video over to a computer. Read chapter 6 to find out more about video media, and chapter 7 to learn about converting old files that will no longer open, if this is an issue. If you read chapter 6, you will find that video is a complex undertaking. You will need to find a method that works for you based on your equipment and needs, and must become familiar with the basics of digital video by reading some books on the subject. Anything older than VHS (Betamax tapes and old film reels) is probably a better project for a video expert rather than a home archivist; consider outsourcing these difficult media projects unless you are committed to spending a lot of time learning about the best way to work with these media.

Digitizing Audio

Audio falls into its own special category. It is certainly not as complicated as video, and most people have at least some basic equipment to assist them in audio conversion (like an old tape deck). You should start by watching some video tutorials or reading about the conversion process before getting started. You should also consider downloading Audacity, a free audio editor discussed in chapters 4 and 6, which is a great tool for amateur audio archivists. Numerous audio editing programs are also available for purchase, and the most highly rated run about $50.[10] These tools are much more approachable than video editors for many people because the buttons and images are similar to what you find on your iPod or old tape deck: stop, play, fast-forward, and record all use the same icons they have for years.

Once audio recording software is ready to go, the main issue is getting the audio to the computer; you will need a handful of specific items for that process. For directions on transferring digital audio (such as voicemails and audio clips) from your phone to your computer using Audacity, see chapter 4. For older media, you will need to own or find the old equipment that the media plays on before you can transfer. Beyond being in working condition, the equipment must also have a 3.5 mm or ¼" headphone jack. You can still find tape decks for cassettes online, and may be able to find used equipment cheap at thrift and second-hand stores. Vinyl is making a resurgence, and you can now find a decent new record player in most electronics departments. With the working equipment, your computer, and audio software, all you will need is a cable with stereo plugs on each end—either 3.5 mm or ¼" depending on the equipment your computer ports. These cables should be available from most electronics departments or online.

With everything set up and connected, you will be ready to convert your audio, and that is where things get complicated. The steps will differ depending on your computer, operating system, and software. This is the point where tutorials should fill in the gaps. While there are books about in-depth audio editing, they are unlikely to be as helpful as online tutorials because the process should be short. Search online for videos using your operating system if you cannot find any with a general search. Terms like "convert cassette to digital Windows 7 tutorial" will help you find what you need.

TIP: If your old media is on cassette or 8-track and you can get the audio from any other source, you should do so. Cassette and 8-track are notoriously poor-quality audio recordings and are only a good option as a master audio version if they are the only copy of a recording.

Old Digital Media

If the media form you own has not been mentioned yet, it may fall into a category of odd things that are digital or semi-digital, but not easily transferrable. For example, the files on your Zip disks or old Commodore cassettes are technically digitally readable, but not by anything you have around the house. If you have old floppies and other outdated media, consider using a service to pull off the files, just as you might for other items in this chapter. Unlike video and photos, however, there are fewer companies for this service, and some additional questions to ask in the process. Another option is to find ways to pull the files from the media sources yourself.

If you are interested in trying to access old digital media yourself, you will likely encounter both hardware and software/file access issues. Hardware is pretty straightforward: what do you need to play or use your media? You may need a floppy drive, a Zip drive, or even an old computer. Based on what you have, search for how-to discussions and tutorials online; chances are other people have gone through similar conversion projects. You may be able to find the old hardware you need by visiting thrift stores and watching online auctions. Some drives and media readers are also still available for purchase. For example, you can purchase an Iomega Zip disk drive on Amazon that will plug in to your computer through USB. Once you find the right equipment, you will hopefully find working, uncorrupted files. Depending on what you find, you may need to look for software conversion tools to view and open old file formats. Information on this process was covered in chapter 7.

If you decide to use a business, get quotes from a few so that you can compare prices. If you live in a metropolitan area, check for local options: computer repair stores sometimes offer this service. If you do not have a local option, companies like Retro Floppy (www.retrofloppy.com) and Data Recovery Masters (www.datarecoverymasters.com) offer this service. Using the service will involve packing and mailing your materials to the company. If you select this option, consider the following questions when talking with companies:

- What media will my files be on when they are returned to me? CD, online download, flash drive, or something else?
- If files on my media are no longer readable by most modern machines, can you convert them to usable file types? If so, how do you choose which types? Will I receive both the legacy file format and converted file format?
- Is there a bulk discount?
- Based on what I have, how long will my project take?

MOVING FORWARD

This chapter has hopefully provided an understanding of the steps involved and your options for digitizing your remaining physical media. Digital media will continue to be created as well—but hopefully, previous chapters have prepared you for the basics of managing those items, too. Whether it is video and photographs, or text messages and old short stories, you can save and protect them using the planning and technological tools outlined in this book. Moving forward, this may seem very intimidating, but managing your digital life successfully is possible.

NOTES

1. Aimee Baldridge, *Organize Your Digital Life: How to Store Your Photographs, Music, Videos, and Personal Documents in a Digital World* (Washington, DC: National Geographic, 2009).

2. Jon Ippolito, "Death by Technology," *Re-Collection: Art, New Media, and Social Memory*, edited by Richard Rinehart and Jon Ippolito (Cambridge, MA: MIT Press, 2014), 32.

3. Catherine C. Marshall, "How People Manage Personal Information over a Lifetime," *Personal Information Management*, edited by William Jones and Jaime Teevan (Seattle: University of Washington Press, 2007), 57–75.

4. Richard Rinehart and Jon Ippolito, *Re-Collection: Art, New Media, and Social Memory* (Cambridge, MA: MIT Press, 2014).

5. Sarah Kim, "The Results of One Scholar's Survey: What Are Your Plans for Your Personal Digital Archives?" *Perspectives on Personal Digital Archiving*, March 2013, http://www .digitalpreservation.gov/documents/ebookpdf_march18.pdf.

6. Baldridge, *Organize Your Digital Life*.

7. Ibid.

8. Barry Wheeler, "What Image Resolution Should I Use?" *Perspectives on Personal Digital Archiving*, March 2013, http://www.digitalpreservation.gov/documents/ebookpdf_march18. pdf.

9. Baldridge, *Organize Your Digital Life*.

10. "Audio Editing Software Reviews," Top Ten Reviews: 2016 Best, accessed September 1, 2016, http://www.toptenreviews.com/software/multimedia/best-audio-editing-software/.

Appendix A

Blank Forms

VALUE ASSESSMENT

MY STUFF	N/A: Does not apply to me.	If I lost this forever, I might not notice or I wouldn't care.	If I lost this forever, it would be inconvenient but I'd be fine.	If I lost this forever, it would ruin my day but I would manage.	If I lost this forever, I would be sad about it for a long while.	If I lost this forever, I would be devastated.
Photos of family and friends						
Emails						
Texts/SMS						
Personal writing (journal, fiction, etc.)						
Bills or financial records						
Medical records						
Scanned letters or correspondence						
Music						
Family videos						
Legal documents						
Saved voicemails						
Genealogy research						
Other:						
Other:						
Other:						
Other:						

LOCATION LIST

DIGITAL ITEMS	LOCATIONS
Photos of family and friends	
Emails	
Texts/SMS	
Personal writing (journal, fiction, etc.)	
Bills or financial records	
Medical records	
Scanned letters or correspondence	
Music	
Family videos	
Legal documents	
Saved voicemails	
Genealogy research	
Other:	
Other:	
Other:	
Other:	

INVENTORY FOR DIGITIZATION

Description (e.g., "photo shoebox #1")	Type (e.g., photos, VHS)	Volume (how many items)	Location (e.g., hall closet, attic)	Importance (1–5)	Keep after digitized? (Yes/No)

Appendix B

A Short Annotated Bibliography
of Personal Digital Archiving Resources

Baldridge, Aimee. *Organize Your Digital Life: How to Store Your Photographs, Music, Videos, and Personal Documents in a Digital World.* Washington, DC: National Geographic, 2009. *Organize Your Digital Life* is a how-to guide with lots of step-by-step instructions. Much of the focus is geared toward digitizing non-digital media rather than dealing with already digital items. It covers basic tips and steps, as well as intermediate how-to explanations like how to install an internal hard drive or set up network-attached storage. Many of the suggested archive options rely on optical media and keeping static extra copies of your files in a "remote location," which was certainly the standard before the advent of cheaper cloud and external storage. Despite some of the recommendations being already out of date six years after publication, the steps for digitizing are still relevant and will be useful, especially to those with large amounts of undigitized materials.

Duffy, Jill E. *Get Organized: How to Clean Up Your Messy Digital Life.* PC Magazine, 2013. As the title implies, *Get Organized* is focused primarily on the organization of your digital life. While it does offer many practical tips for improving management of email and files, it also makes strong recommendations for applications and productivity practices that may not work for everyone. The author admits in the introduction that many of the recommendations evolved from her own practices, which results in specific needs being addressed. Suggestions for search improvement, file naming, and deleting files are relevant to all computer users, as are the monthly and annual cleanup checklists in the appendix. Things like cleaning up the desktop, setting up user accounts, and organizing Windows 8 tiles are also covered. It is a useful book for people with straightforward digital needs who want tips for better organization and efficiency.

Hawkins, Donald T., ed. *Personal Archiving: Preserving Our Digital Heritage.* Medford, NJ: Information Today, 2013. This collection, edited by Hawkins but including many contributors, takes a practical but also an academic approach to personal digital archiving. Writers from the Library of Congress, Microsoft Research, Penn State Libraries, and other well-known institutions share both specific workflows and projects as well as more theoretical discussions on the importance of personal digital archiving. The book suggests that it is for family archivists/historians, academic researchers, librarians, historians, and public officials—a wide range of individuals. It will likely appeal to all of them in some fashion. From chapters discussing specific software to help personal archivists, to the challenges of specific individuals who approached projects in different ways, this book makes it is possible to get an overview of this complex topic. It is a useful resource for anyone seeking to gain a solid foundation on personal digital archiving.

Jones, William, and Jaime Teevan, eds. *Personal Information Management.* Seattle: University of Washington Press, 2007. This book takes an academic/research-based approach in discussing how individuals manage personal information, including digital information. With specific examples of people and scenarios throughout, it addresses common problems as well as some more uncommon personal information needs (such as attempting to capture your entire life on video). Examples are used to illustrate the individuality of personal information management and how current tools and workflows do not fit the needs of some people. Though it is now almost ten years old, it offers a strong theoretical framework for how people manage their own information and the challenges involved for those individuals, information professionals, and researchers.

Rinehart, Richard, and Jon Ippolito. *Re-Collection: Art, New Media, and Social Memory.* Cambridge, MA: MIT Press, 2014. *Re-Collection* looks at personal digital archiving within the scope of art and culture in historical context. Chapters are grouped into sections based on the different risks associated with new and digital media: death by technology, death by institution, and death by law. The authors discuss numerous issues related to how art and culture are formed in the digital age, and what the implications are for the future. The final chapter, "Only You Can Prevent the End of History," suggests specific steps for different stakeholders (curators, lawyers, creators, etc.) to "Future-proof contemporary culture." Though there are some concrete suggestions in this final chapter, the book is not meant to be a how-to guide. Instead, it is intended to further the discussion on the importance of the continuation of social memory and how we can all contribute.

Index

Page references for figures are italicized.

About the Author

Melody Condron is a librarian at the University of Houston Libraries, where she oversees a number of quality-control activities within the library collections and records. She is professionally active in the American Library Association (ALA), Library Information and Technology Association (LITA), and Texas Library Association (TLA). She has previously worked at the Montana State Library and the Lincoln County Public Libraries in Montana. She has presented at the Personal Digital Archiving conference, LITA Forum, and other conferences, and has instructed a month-long course on personal digital archiving for librarians through LITA.

In addition to her professional work in information organization, Melody is also her family archivist and genealogist, and has digitized over 20,000 family photographs. She holds a BA in communications from Penn State, The Behrend College, and an MLS from the University of North Texas. Her personal interests include urban homesteading/gardening, animal rescue, genealogy, cooking, gaming, competitive Scrabble, and Lego. Melody is originally from Erie, Pennsylvania.